Nonlinear Systems and Complexity

Volume 25

Series editor

Albert C. J. Luo
Southern Illinois University
Edwardsville, IL, USA

Nonlinear Systems and Complexity provides a place to systematically summarize recent developments, applications, and overall advance in all aspects of nonlinearity, chaos, and complexity as part of the established research literature, beyond the novel and recent findings published in primary journals. The aims of the book series are to publish theories and techniques in nonlinear systems and complexity; stimulate more research interest on nonlinearity, synchronization, and complexity in nonlinear science; and fast-scatter the new knowledge to scientists, engineers, and students in the corresponding fields. Books in this series will focus on the recent developments, findings and progress on theories, principles, methodology, computational techniques in nonlinear systems and mathematics with engineering applications. The Series establishes highly relevant monographs on wide ranging topics covering fundamental advances and new applications in the field. Topical areas include, but are not limited to: Nonlinear dynamics Complexity, nonlinearity, and chaos Computational methods for nonlinear systems Stability, bifurcation, chaos and fractals in engineering Nonlinear chemical and biological phenomena Fractional dynamics and applications Discontinuity, synchronization and control.

More information about this series at http://www.springer.com/series/11433

Constantin Milici • Gheorghe Drăgănescu
J. Tenreiro Machado

Introduction to Fractional Differential Equations

Springer

Constantin Milici
Department of Mathematics
Polytechnic University of Timişoara
Timişoara, Romania

Gheorghe Drăgănescu
Research Center in Theoretical Physics
West University of Timişoara
Timişoara, Romania

J. Tenreiro Machado
Department of Electrical Engineering
Polytechnic of Porto
Porto, Portugal

ISSN 2195-9994 ISSN 2196-0003 (electronic)
Nonlinear Systems and Complexity
ISBN 978-3-030-00894-9 ISBN 978-3-030-00895-6 (eBook)
https://doi.org/10.1007/978-3-030-00895-6

Library of Congress Control Number: 2018956771

This Springer imprint is published by the registered company Springer Nature Switzerland AG
The registered company address is: Gewerbestrasse 11, 6330 Cham, Switzerland

Preface

The concept of fractional derivative (Fractional derivative (FD)) was introduced after 1695 as a simply academic generalization of integer derivative. An FD generalizes the order of differentiation from positive integers Set of natural numbers ($\mathbb{N}$) to real Set of real numbers ($\mathbb{R}$), or even to complex Set of complex numbers ($\mathbb{C}$) numbers. A detailed presentation of the old historical steps of fractional calculus (Fractional calculus (FC)) with references is presented in the papers [45, 50].

New sporadic investigation in this field was published after 1930. A deep research was carried out by E.L. Post [31, 52]. More recently we find a series of definitions of the FD [7]. A presentation of the recent history steps can be found in the papers [46–48, 51]. It was published in a series of monographic books as [6, 21, 22, 26, 37, 38].

In the last two decades, it was established that a series of phenomena can be studied in terms of FC. It was established that the rheologic properties [34] of some polymers can be expressed with the aid of fractional differential models [1, 2, 10, 15, 20, 26, 28, 29, 36]. Fractional phenomena were established as the damping phenomena in the high-density polyurethane foams [42], nuclear reactor dynamics [35], thermoelasticity [33], mechanical vibrations [8], or biological tissues [5, 19]. An analysis of the integer and fractional entropy is performed in [44].

It was also experimentally verified fractional diffusion phenomena and fractional electrolytic coating process [4, 30, 32]. This confirms that the roughness of the electrolytic metallic coatings has a fractal structure and can be described in terms of fractional diffusion [3] and stochastic differential process [22].

Quantum fractional differential models were studied also in the book of R. Herrmann [11] or in the papers of Saxena [39], Xiao et al. [55], and Yang et al. [53].

Recently, several fractional devices were developed, containing electrical, thermal, and mechanical components [12, 14]. Also, a series of fractional dynamics experiments are presented in the book of Biswas et al. [4].

It was designed and has been achieved experimentally a series of fractional systems [9, 16, 56] involving control [18, 49] and fractional controllers [43]. It was introduced a series of identification methods for fractional dynamic systems [13]. Fractional wavelet bases [54] in the field of signal processing were discussed [40]. It is important to tell that FD of a periodic function is also periodic [27].

This review of the possible applications of FC in the real world justifies the necessity of its extensive study. The aim of this book is to introduce a series of problems and methods insufficiently discussed in the field of FC.

A series of examples based on symbolic computation, written in Maple© and Mathematica©, are presented. The reader can find other useful applications for the case of integer order systems in the book of Inna Shingareva and C. Lizárraga-Celaya [41], which can be extended to the case of fractional calculus, or the book of problems [17].

This book is organized in six chapters.

Chapter 1 This chapter presents the most important special functions involved in FC. A special attention is devoted to the Gamma function, used in FC calculations. Other special functions such as the Euler, Beta, and Mittag-Leffler functions (Mittag-Leffler function (MLF)) are also introduced.

Chapter 2 This chapter introduces the fractional integral (Fractional integral (FI)) and FD, in the sense of Riemann–Liouville. The properties of these fractional operators are discussed.

Chapter 3 This chapter is devoted to the use of the Laplace transform (Laplace transform (LT)) in FC, because the Riemann–Liouville FI and FD allow the derivation of closed-form solutions with the aid of the LT method.

Chapter 4 This chapter is devoted to the nonlinear fractional differential equations (Fractional differential equation (FDE)). The text discusses the generalization of the methods used to solve the integer order differential equations. In this line of thought, approaches such as the Picard array, Adomian decomposition, and perturbation methods are analyzed.

Chapter 5 In this chapter several classical models are generalized in the perspective of FC. First, the fractional integral sine and cosine functions are formulated, and the corresponding spirals, projected on the plane, sphere, and cone, are illustrated. Second, the fractional generalizations of the Lane-Emden, Hermite, Legendre, and Bessel equations are studied. It is also discussed the power series method for their solution.

Chapter 6 This chapter extends standard numerical algorithms, namely, the least squares, Galerkin, and Euler methods, that are applied to FDE. A special attention is given the Runge–Kutta (Runge–Kutta (RK)) method for fractional equations and systems. The algorithms are applied to a series of FDE and systems of FDE.

Moreover, a generalized method, based on decomposition and LT, is presented [23–25].

These new algorithms are illustrated by means of the Maple and Mathematica software packages.

The content of this book is addressed to large category of readers, working in the fields of fundamental and applied mathematics, theoretical and experimental physics, experimental engineering, and others. The authors hope that the material included in the book help researchers to enter to the huge emerging scientific area of FC and its applications.

Timişoara, Romania — Constantin Milici
Timişoara, Romania — Gheorghe Drăgănescu
Porto, Portugal — J. Tenreiro Machado

References

1. Atanackovič, T. M., Pilipovič, S., Stankovič, B., & Zorica, D. (2014). *Fractional calculus with applications in mechanics: Wave propagation, impact and variational principles*. London: Wiley-ISTE.
2. Atanackovič, T. M., Pilipovič, S., Stankovič, B., & Zorica, D. (2014). *Fractional calculus with applications in mechanics: Vibrations and diffusion processes*. London: Wiley-ISTE.
3. Barabási, A.-L., & Stanley, H. E. (2002). *Fractal concepts in surface growth* (2nd ed.). Cambridge: Cambridge University Press.
4. Biswas, K., Bohannan, G., Caponetto, R., Mendes Lopes, A., & Tenreiro Machado, J. A. (2017). *Fractional-order devices. SpringerBriefs in nonlinear circuits*. New York: Springer.
5. Bueno-Orovio, A., Kay, D., Grau, V., Rodriguez, B., & Burrage, K. (2014). Fractional diffusion models of cardiac electrical propagation: Role of structural heterogeneity in dispersion of repolarization. *Journal of the Royal Society Interface, 11*, 20140352.
6. Das, S. (2011). *Functional fractional calculus*. Berlin: Springer.
7. de Oliveira, E. C., & Tenreiro Machado, J. A. (2014). A review of definitions for fractional derivatives and integral. *Mathematical Problems in Engineering, 2014*, Article ID 238459.
8. Deü, J.-F., & Matignon, D. (2010). Simulation of fractionally damped mechanical systems by means of a Newmark-diffusive scheme. *Computers & Mathematics with Applications, 59*, 1745–1753.
9. Diethelm, K., Ford, N. J., & Freed, A. D. (2002). Simulation of fractionally damped mechanical systems by means of a Newmark-diffusive scheme. *Nonlinear Dynamics, 59*, 3–22.

10. Drăgănescu, G. E. (2006). Application of a variational iteration method to linear and nonlinear viscoelastic models with fractional derivatives. *Journal of Mathematical Physics, 47*, 082902.
11. Herrmann, R. (2011). *Fractional calculus. An introduction for physicists*. Singapore: World Scientific Publishing Company.
12. Hughes, R. N. (1992). *Design and analysis of electrical circuits that produce fractional-order differentiation*. Master's thesis, Air University, Wright-Patterson Air Force Base, Ohio. AFIT/GE/ENG/92M-05.
13. Jalloul, A., Jelassi, K., & Trigeassou, J.-C. (2013). A comparative study of identification techniques for fractional models. *International Journal of Electrical & Computer Engineering, 3*, 186–196.
14. Jesus, I. S., Tenreiro Machado, J. A., & Boaventura Cunha, J. (2008). Fractional electrical impedances in botanical elements. *Journal of Vibration and Control, 14*, 1389–1402.
15. Kilbas, A. A., Srivastava, H. M., & Trujillo, J. J. (2006). *Theory and applications of fractional differential equations*. Amsterdam: Elsevier.
16. Klafter, J., Lim, S. C., & Metzler, R. (2011). *Fractional dynamics: Recent advances*. Singapore: World Scientific Publishing Company.
17. Krasnov, M. L., Kiselyov, A. I., & Makarenko, G. I. (1981). *A book of problems in ordinary differential equations*. Moscow: Mir Publishers.
18. Luo, Y., & Chen, Y. (2012). *Fractional order motion controls*. Chichester: John Wiley & Sons.
19. Magin, R. L. (2010). Fractional calculus models of complex dynamics in biological tissuese. *Computers & Mathematics with Applications, 59*, 1586–1593.
20. Mainardi, F., & Spada, G. (2011). Creep, relaxation and viscosity properties for basic fractional models in rheology. *The European Physical Journal Special Topics, 193*, 133–160.
21. Malinowska, A. B., Odzijewicz, T., & Torres, D. F. M. (2015). *Advanced methods in the fractional calculus of variations. SpringerBriefs in applied sciences and technology*. Heidelberg: Springer.
22. Meerschaert, M. M., & Sikorskii, A. (2012). *Stochastic models for fractional calculus. De Gruyter studies in mathematics*. Heidelberg: Walter de Gruyter GmbH & Co.
23. Milici, C., & Drăgănescu, G. (2014). *A method for solve the nonlinear fractional differential equations*. Saarbrücken: Lambert Academic Publishing.
24. Milici, C., & Drăgănescu, G. (2015). *New methods and problems in fractional calculus*. Saarbrücken: Lambert Academic Publishing.
25. Milici, C., & Drăgănescu, G. (2017). *Introduction to fractional calculus*. Saarbrücken: Lambert Academic Publishing.
26. Oldham, K. B., & Spanier, J. (1973). *The fractional calculus: Theory and applications of differentiation and integration to arbitrary order*. San Diego: Academic Press.

27. Ortigueira, M. D., Tenreiro Machado, J., & Trujillo, J. J. (2017). Fractional derivatives and periodic functions. *International Journal of Dynamics and Control, 5*, 72–78.
28. Oustaloup, A. (2014). *Diversity and non-integer differentiation for system dynamics*. Hoboken: Wiley-ISTE.
29. Petráš, I. (2011). *Fractional-order nonlinear systems modeling, analysis and simulation*. Beijing/Heidelberg: Higher Education Press/Springer.
30. Podlubny, I. (1998). *Fractional differential equations: An introduction to fractional derivatives, fractional differential equations, to methods of their solution and some of their applications. Mathematics in science and engineering*. San Diego: Academic Press.
31. Post, E. L. (1930). Generalized differentiation. *Transactions of the American Mathematical Society, 32*, 723–781.
32. Povstenko, Y. (2015). *Linear fractional diffusion-wave equation for scientists and engineers*. Cham: Birkhäuser.
33. Povstenko, Y. (2015). *Fractional thermoelasticity*. Cham: Springer.
34. Pritchard, R. H., & Terentjev, E. M. (2017). Oscillations and damping in the fractional Maxwell materials. *Journal of Rheology, 61*, 187–203.
35. Ray, S. S. (2016). *Fractional calculus with applications for nuclear reactor dynamics*. Boca Raton: CRC Press.
36. Rossikhin, Yu. A., & Shitikova, M. V. (2004). Analysis of the viscoelastic rod dynamics via models involving fractional derivatives or operators of two different orders. *The Shock and Vibration Digest, 36*, 3–26.
37. Sabatier, J., Agrawal, O. P., & Tenreiro Machado, J. A. (Eds.). (2007). *Advances in fractional calculus theoretical developments and applications in physics and engineering*. Dordrecht: Springer.
38. Samko, S. G., Kilbas, A. A., & Marichev, O. I. (Eds.). (1993). *Fractional integrals and derivatives: Theory and applications*. Yverdon: Gordon and Breach Science Publishers.
39. Saxena, R. K., Saxena, R., & Kalla, S. L. (2010). Solution of spacetime fractional Schrödinger equation occurring in quantum mechanics. *Fractional Calculus & Applied Analysis, 13*, 177–190.
40. Sheng, H., Chen, Y. & Qiu, T. (Eds.). (2012). *Fractional processes and fractional-order signal processing: Techniques and applications*. London: Springer.
41. Shingareva, I. K., & Lizárraga-Celaya, C. (Eds.). (2011). *Solving nonlinear partial differential equations with maple and mathematica*. Wien: Springer.
42. Singh, R., Davies, P., & Bajaj, A. K. (2003). Identification of nonlinear and viscoelastic properties of flexible polyurethane foam. *Nonlinear Dynamics, 34*, 319–346.
43. Tenreiro Machado, J. (2001). Discrete-time fractional-order controllers. *Fractional Calculus & Applied Analysis, 4*, 47–66.
44. Tenreiro Machado, J. A. (2010). Entropy analysis of integer and fractional dynamical systems. *Nonlinear Dynamics, 62*, 371–378.

45. Tenreiro Machado, J. A., Mainardi, F., & Kiryakova, V. (2010). A poster about the old history of fractional calculus. *Fractional Calculus & Applied Analysis, 13*, 447–454.
46. Tenreiro Machado, J., Kiryakova, V., & Mainardi, F. (2011). Recent history of fractional calculus. *Communications in Nonlinear Science and Numerical Simulation, 16*, 1140–1153.
47. Tenreiro Machado, J., Lopes, A. M., Duarte, F. B., Ortigueira, M. D., & Rato, R. T. (2014). Rapsody in fractional. *Fractional Calculus & Applied Analysis, 17*, 1188–1214.
48. Tenreiro Machado, J. A., Mainardi, F., Kiryakova, V., & Atanacković, T. (2016). Fractional calculus: D'où venons-nous? Que sommes-nous? Où allons-nous? *Fractional Calculus & Applied Analysis, 17*, 1074–1104.
49. Valério, D., & Costa, J. S. (2013). *An introduction to fractional control.* London: CRC Press.
50. Valério, D., Tenreiro Machado, J. A., & Kiryakova, V. (2014). Some pioneers of the applications of fractional calculus. *Fractional Calculus & Applied Analysis, 17*, 552–578.
51. Weilbeer, M. (2005). *Efficient numerical methods for fractional differential equations and their analytical background.* PhD thesis, Facultät für Mathematik und Informatik, Technischen Univerisität Braunschweig, Braunschweig.
52. Wheeler, N. (1997). *Construction and physical application of fractional calculus.* Technical report, Reed College Physics Department.
53. Yang, X.-J., Baleanu, D., & Tenreiro Machado, J. A. (2013). Mathematical aspects of the Heisenberg uncertainty principle within local fractional Fourier analysis. *Boundary Value Problems, 131*, 1–16.
54. Yang, X.-J., Baleanu, D., Srivastava, H. M., & Tenreiro Machado, J. A. (2013). On local fractional continuous wavelet transform. *Abstract and Applied Analysis, 2013*, 5.
55. Zhang, X., Wei, C., Liu, Y., & Luo, M. (2014). Fractional corresponding operator in quantum mechanics and applications: A uniform fractional Schrödinger equation in form and fractional quantization methods. *Annals of Physics, 350*, 124–136.
56. Zhou, Y., Ionescu, C., & Tenreiro Machado, J. A. (2015). Fractional dynamics and its applications. *Nonlinear Dynamics, 80*, 1661–1664.

Contents

Acronyms

FC	Fractional calculus
FD	Fractional derivative
FDE	Fractional differential equation
FI	Fractional integral
LT	Laplace transform
RK	Runge–Kutta
RK2	Second order Runge–Kutta
RK4	Fourth order Runge–Kutta
VIM	Variational iteration method

List of Symbols

$E(t, \alpha, a)$	E function
$J_p(t)$	Bessel function of first kind
B	Beta function
erfc	Complementary error function
$\mathbb{C}$	Set of complex numbers
det	Determinant
erf	Error function
γ	Euler's constant
Ei	Exponential integral function
Γ	Gamma function
erfi	Imaginary error function
L	Laplace transform
max	Maximum
min	Minimum
E_α	One parameter Mittag-Leffler function
$\mathbb{N}$	Set of natural numbers
$\mathbb{R}$	Set of real numbers
Re	Real part
Res	Residues
sup	Supremum
$E_{\alpha,\beta}$	Two parameter Mittag-Leffler function

Chapter 1
Special Functions

In this chapter several special functions used in the follow-up of the book are presented briefly. More details about these functions can be found in [1, 3, 4, 8].

1.1 Euler's Function

1.1.1 Gamma Function

We start by considering the Gamma function, or second order Euler[1] integral, denoted $\Gamma(\cdot)$ represented in Fig. 1.1.

Function Gamma function (Γ) is defined as:

$$\Gamma(p) = \int_0^\infty e^{-x} x^{p-1} dx. \tag{1.1}$$

Theorem *Function $\Gamma(p)$ is convergent for $p > 0$.*

Proof The integral can be written as:

$$\Gamma(p) = \int_0^1 e^{-x} x^{p-1} dx + \int_1^\infty e^{-x} x^{p-1} dx = I_1 + I_2,$$

where $I_1 = \displaystyle\int_0^1 e^{-x} x^{p-1} dx$ is convergent.

---

[1] L. Euler (1707–1783).

C. Milici et al., *Introduction to Fractional Differential Equations*, Nonlinear Systems and Complexity 25, https://doi.org/10.1007/978-3-030-00895-6_1

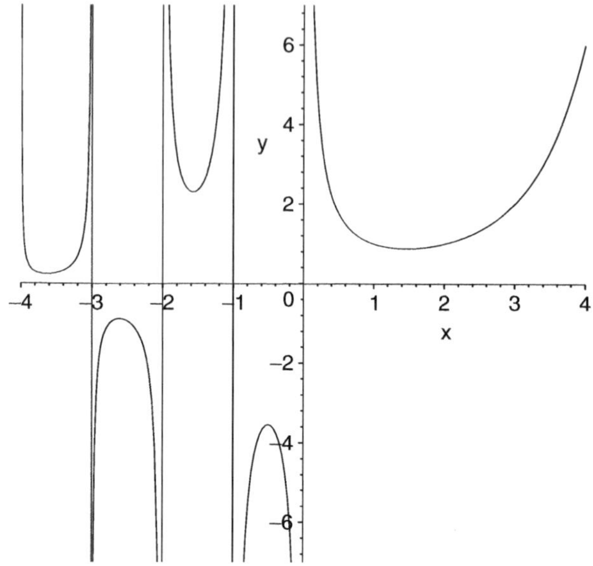

Fig. 1.1 The plot of $y = \Gamma(x)$ function

Since e^{-x} is decreasing on the interval [0, 1], from $x = 0$, we have:

$$\int_0^1 e^{-x} x^{p-1} dx < \int_0^1 x^{p-1} dx = \frac{1}{p}.$$

Moreover, $I_2 = \int_1^{\infty} e^{-x} x^{p-1} dx$ is also convergent. We obtain:

$$1 \leq x \ \Rightarrow \ x^{p-1} e^{-x} \leq e^{-x/2} \ \Leftrightarrow \ x^{p-1} \leq e^{x/2} \ \Leftrightarrow \ \frac{x^{p-1}}{e^{x/2}} \leq 1.$$

Because $\lim_{x \to \infty} \frac{x^{p-1}}{e^{x/2}} = 0$, we have:

$$\int_1^{\infty} e^{-x} x^{p-1} dx \leq \int_1^{\infty} e^{-x/2} dx = 2e^{-1/2}.$$

The integral (1.1) is convergent for $p > 0$ and divergent for $p \leq 0$.
The basic properties of the Gamma function are:

1. The function $\Gamma(p)$ is continuous for $p > 0$.
2. The function $\Gamma(p)$ obeys the property:

$$\Gamma(p+1) = p\,\Gamma(p). \tag{1.2}$$

Proof $\Gamma(p+1) = \int_0^\infty e^{-x}x^p dx = -[e^{-x}x^p]_0^\infty + p\int_0^\infty e^{-x}x^{p-1}dx = p\Gamma(p).$

3. The following relations are also valid:

$$\Gamma(p+n) = (p+n-1)\ldots(p+1)\,p\,\Gamma(p), \tag{1.3}$$

$$\Gamma(1) = 1,$$

$$\Gamma(n+1) = n!,$$

$$\Gamma(0) = +\infty.$$

4. For $p = -n$ it results:

$$\Gamma(-n) = \frac{\Gamma(-n+1)}{-n}$$
$$= \frac{\Gamma(-n+2)}{n(n-1)} = \frac{\Gamma(-n+3)}{n(n-1)(n-2)} = \ldots = (-1)^n\frac{\Gamma(0)}{n!} = (-1)^n\infty.$$

5. Taking account that the Γ function can be written as $\Gamma(p) = \dfrac{\Gamma(p+1)}{p}$, it results that the Γ function can be defined also for negative values of p, in the interval $-1 < p < 0$.

   If $-n < p < -(n-1)$, then from (1.3) it results:

$$\Gamma(p) = \frac{\Gamma(p+n)}{p(p+1)\ldots(p+n-1)}.$$

   Using the substitution $p + n = \alpha$, it results after calculations:

$$\Gamma(\alpha - n) = \frac{(-1)^n\Gamma(\alpha)}{(1-\alpha)(2-\alpha)\ldots(n-\alpha)}.$$

6. Using the identity (1.2) we obtain:

$$\Gamma\left(m+\frac{1}{2}\right) = \Gamma\left[1+\left(m-\frac{1}{2}\right)\right] = \left(m-\frac{1}{2}\right)\Gamma\left(m-\frac{1}{2}\right)$$
$$= \left(m-\frac{1}{2}\right)\left(m-\frac{3}{2}\right)\Gamma\left(m-\frac{1}{2}\right)$$
$$= \left(m-\frac{1}{2}\right)\left(m-\frac{3}{2}\right)\ldots\frac{5}{2}\frac{3}{2}\frac{1}{2}\Gamma\left(\frac{1}{2}\right),$$

or:

$$\Gamma\left(m+\frac{1}{2}\right)=\frac{(2m-1)!}{2^m}\Gamma\left(\frac{1}{2}\right)=\frac{(2m)!}{m!2^{2m}}\Gamma\left(\frac{1}{2}\right).$$

7. We can prove the identity [4]:

$$\Gamma(p)=\int_0^1\left(\ln\frac{1}{y}\right)^{p-1}dy$$

8. The following particular values for Γ function can be useful for calculation purposes:

$$\Gamma\left(\frac{1}{2}\right)=\sqrt{\pi},$$

$$\Gamma\left(-\frac{1}{2}\right)=-2\sqrt{\pi},$$

$$\Gamma\left(\frac{3}{2}\right)=\Gamma\left(1+\frac{1}{2}\right)=\frac{1}{2}\Gamma\left(\frac{1}{2}\right)=\frac{1}{2}\sqrt{\pi},$$

$$\Gamma\left(\frac{5}{2}\right)=\Gamma\left(2+\frac{1}{2}\right)=\frac{4!\Gamma\left(\frac{1}{2}\right)}{2!\,2^4}=\frac{3}{4}\sqrt{\pi},$$

$$\Gamma\left(\frac{1}{3}\right)=2.678938 \qquad \Gamma\left(m+\frac{1}{3}\right)=\frac{1\,4\,\ldots\,(3m-2)}{3^m}\Gamma\left(\frac{1}{3}\right),$$

$$\Gamma\left(\frac{2}{3}\right)=1.354118 \qquad \Gamma\left(m+\frac{2}{3}\right)=\frac{2\,5\,\ldots\,(3m-1)}{3^m}\Gamma\left(\frac{2}{3}\right),$$

$$\Gamma\left(\frac{1}{4}\right)=3.625600 \qquad \Gamma\left(m+\frac{1}{4}\right)=\frac{1\,5\,\ldots\,(4m-3)}{4^m}\Gamma\left(\frac{1}{4}\right),$$

$$\Gamma\left(\frac{3}{4}\right)=1.225417 \qquad \Gamma\left(m+\frac{3}{4}\right)=\frac{3\,7\,\ldots\,(4m-1)}{4^m}\Gamma\left(\frac{3}{4}\right).$$

9.

$$\frac{\Gamma(p+1)}{\Gamma(q+1)\Gamma(p-q+1)}=\binom{p}{q},$$

10. The Gauss'[2] formula is:

$$\Gamma(p) = \frac{1}{p} \prod_{k=1}^{\infty} \left(1 + \frac{1}{k}\right)^{p} \left(1 + \frac{p}{k}\right)^{-1}.$$

Proof We express e^{-x} as:

$$e^{-x} = \lim_{k\to\infty} \left(1 - \frac{x}{k}\right)^{k}.$$

Then, we obtain:

$$\Gamma(p) = \int_0^{\infty} e^{-x} x^{p-1} dx = \lim_{k\to\infty} \int_0^{k} \left(1 - \frac{x}{k}\right)^{k} x^{p-1} dx.$$

For $x = t\,k \quad \Rightarrow dx = k\,dt$, resulting:

$$\Gamma(p) = \lim_{k\to\infty} k^p \int_0^1 (1-t)^k t^{p-1} dt.$$

Integrating by parts we obtain:

$$\frac{1}{p} \int_0^1 (1-t)^k dt^p = \frac{1}{p} \Big[(1-t)^k t^p\Big]_0^1 - \frac{1}{p} \int_0^1 t^p d(1-t)^k = \frac{k}{p} \int_0^1 (1-t)^{k-1} t^p dt.$$

Repeating this operation it follows:

$$\Gamma(p) = \lim_{k\to\infty} \frac{k^p k!}{p(p+1)\ldots(p+k)}.$$

But:

$$\lim_{k\to\infty} \frac{(k+1)^p}{k^p} = 1.$$

From:

$$\Gamma(p) = \frac{1}{p} \lim_{k\to\infty} \frac{1}{(1+p)(1+p/2)\ldots(1+p/k)} \frac{2^p 3^p \ldots (k+1)^p}{1^p 2^p \ldots k^p},$$

it follows:

$$\Gamma(p) = \frac{1}{p} \prod_{k=1}^{\infty} \frac{1}{(1+p/k)} \frac{(k+1)^p}{k^p} = \frac{1}{p} \prod_{k=1}^{\infty} \left(1 + \frac{1}{k}\right)^{p} \left(1 + \frac{p}{k}\right)^{-1},$$

[2]J.C.F. Gauss (1777–1855).

excepting the values $\mathrm{Re}(p) = 0, -1, -2, \ldots$.

11. The Weierstrass[3] form of the Gamma function is:

$$\frac{1}{\Gamma(p)} = pe^{\gamma p} \prod_{k=1}^{\infty} \left(1 + \frac{1}{k}\right) e^{-p/k}$$

valid, excepting the values $\mathrm{Re}(p) = 0, \ -1, \ -2, \ldots, -n, \ldots,$ where Real part (Re) represents the real part and $n \in \mathbb{N}$.

Here the symbol Euler's constant (γ) represents the Euler's constant, given by:

$$\gamma = \lim_{p \to \infty} \left(\sum_{k=1}^{p} \frac{1}{k} - \ln p \right) = 0.577215663\ldots$$

Proof The Gauss' formula can be written as:

$$\Gamma(p) = \lim_{k \to \infty} \frac{k^p k!}{p(p+1)\ldots(p+k)},$$

excepting for $\mathrm{Re}(p) = 0, -1, -2, \ldots$. We can write also:

$$k^p = e^{p \ln k} = e^{p(\ln k - 1 - 1/2 - \ldots - 1/k)} e^{p + p/2 + \ldots + p/k},$$

$$\Gamma_k(t) = \frac{e^{p(\ln k - 1 - 1/2 - \ldots - 1/k)} e^{p + p/2 + \ldots + p/k}}{p(1+p)(1+p/2)\ldots(1+p/k)},$$

$$\Gamma_k(t) = \exp\left[p \left(\ln k - 1 - \frac{1}{2} - \ldots - \frac{1}{k} \right) \right] \frac{e^p}{1+p} \frac{e^{\frac{p}{2}}}{1+\frac{p}{2}} \cdots \frac{e^{\frac{p}{k}}}{1+\frac{p}{k}},$$

and finally:

$$\frac{1}{\Gamma(p)} = \lim_{k \to \infty} \frac{1}{\Gamma_k(t)} = e^{p\gamma} \lim_{k \to \infty} e^{-p}(1+p)e^{-\frac{p}{2}} \ldots e^{-\frac{p}{k}} \left(1 + \frac{p}{k}\right)$$

$$= pe^{\gamma p} \prod_{k=1}^{\infty} \left(1 + \frac{p}{k}\right) e^{-\frac{p}{k}}.$$

12. *Reflection* property of the Gamma function is given by:

$$\Gamma(p)\,\Gamma(1-p) = \frac{\pi}{\sin(\pi p)}, \quad 0 \le p \le 1.$$

[3]K.T.W. Weierstrass (1815–1897).

Proof From the Weierstrass' formula we have:

$$\frac{1}{\Gamma(p)}\frac{1}{\Gamma(-p)} = -p^2 e^{\gamma_p} e^{-\gamma_p} \prod_{k=1}^{\infty}\left(1+\frac{p}{k}\right)e^{-p/k}\left(1-\frac{p}{k}\right)e^{p/k}$$
$$= -p^2 \prod_{k=1}^{\infty}\left(1-\frac{p^2}{k^2}\right).$$

But $\Gamma(-p) = \dfrac{\Gamma(1-p)}{p}$. It results:

$$\frac{1}{\Gamma(p)}\frac{1}{\Gamma(1-p)} = p\prod_{k=1}^{\infty}\left(1-\frac{p^2}{k^2}\right).$$

Because $\sin(n\pi) = \pi n \prod_{k=1}^{\infty}\left(1-\frac{p^2}{k^2}\right)$ it results the reflection property.

Another Proof

$$\Gamma(p)\Gamma(1-p) = \int_0^{\infty} e^{-t}t^{p-1}dt \int_0^{\infty} e^{-s}s^{1-p-1}ds$$
$$= \int_0^{\infty}\int_0^{\infty} e^{-(t+s)}\left(\frac{t}{s}\right)^p t^{-1}dtds.$$

Using the changes of variables

$$t+s=u, \quad \frac{t}{s}=v,$$

and the residue theory it can be obtained:

$$\Gamma(p)\Gamma(1-p) = \int_0^1 \frac{v^p}{1+v}dv = \frac{\pi}{\sin(p\pi)}.$$

Figure 1.2 depicts $\frac{1}{\Gamma(x)}$.

1.1.2 Beta Function

Here we consider the Beta function, denoted Beta function (B).

The Beta function, or the first order Euler function, can be defined as [2, 7]:

$$B(p,q) = \int_0^1 x^{p-1}(1-x)^{q-1}dx,$$

where $\text{Re}(p) > 0$ and $\text{Re}(q) > 0$.

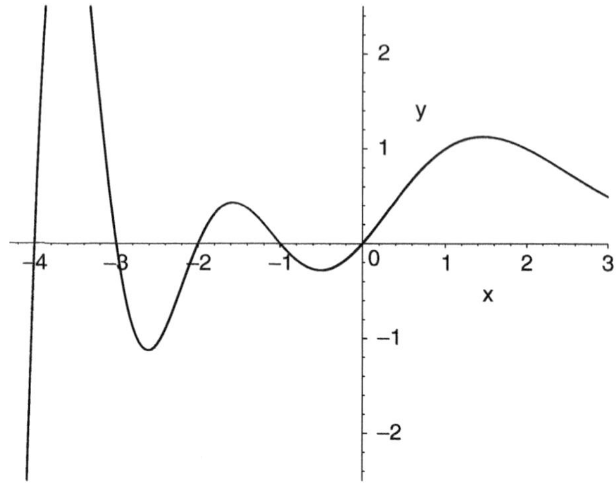

Fig. 1.2 The plot of $y = 1/\Gamma(x)$ function

In the following we will enumerate the basic properties of the Beta function:

1. For every $p > 0$ and $q > 0$, we have:

$$B(p,q) = B(q,p).$$

2. For every $p > 0$ and $q > 1$, the Beta function B satisfies the property:

$$B(p,q) = \frac{q-1}{p+q-1} B(p,q-1).$$

Proof

$$B(p,q) = \int_0^1 x^{p-1}(1-x)^{q-1}\,dx,$$

$$x^p(1-x)^{q-2} = x^{p-1}(1-x)^{q-2} - x^{p-1}(1-x)^{q-1},$$

$$\begin{aligned} B(p,q) &= \int_0^1 (1-x)^{q-1} d\frac{x^p}{p} = \frac{x^p(1-x)^{q-1}}{p}\Big|_0^1 + \frac{q-1}{p}\int_0^1 x^p(1-x)^{q-2}dx \\ &= \frac{q-1}{p}\int_0^1 x^{p-1}(1-x)^{q-2}dx - \frac{q-1}{p}\int_0^1 x^{p-1}(1-x)^{q-1}dx \\ &= \frac{q-1}{p}B(p,q-1) - \frac{q-1}{p}B(p,q). \end{aligned}$$

3. For every $p > 0$ and $q > 0$, it is valid the identity:

$$B(p,q) = \frac{\Gamma(p)\Gamma(q)}{\Gamma(p+q)}.$$

Proof The product $\Gamma(p)\Gamma(q)$ can be written as:

$$\Gamma(p)\Gamma(q) = \int_0^\infty e^{-t}t^{p-1}dt \int_0^\infty e^{-s}s^{p-1}ds = \int_0^\infty \int_0^\infty e^{-(t+s)}t^{p-1}s^{q-1}dtds,$$

$$\Gamma(p+q) = \int_0^\infty \int_0^\infty e^{-(t+s)}t^{p-1}s^{q-1}\, dt\, ds.$$

We use the notation $t + s = x$, for $0 < t < \infty$ and $0 < t < \infty$.
The Jacobian is

$$\frac{D[t,s]}{D[x,y]} = \begin{Vmatrix} y & x \\ 1-y & -x \end{Vmatrix} = -xy - x + xy = -x,$$

resulting:

$$dtds = \left|\frac{D[t,s]}{D[x,y]}\right| dxdy = xdxdy,$$

$$\Gamma(p)\Gamma(q) = \int_0^1 \int_0^1 e^{-x}(xy)^{p-1}x^{q-1}(1-y)^{q-1}xdxdy$$
$$= \int_0^\infty e^{-x}x^{p+q-1}dx \int_0^1 y^{p-1}(1-y)^{q-1}dy,$$
$$\Gamma(p)\Gamma(q) = \Gamma(p+q)B(p,q).$$

4. For every $p > 0$, and for the natural number n, it can be proved

$$B(p,n) = B(n,p) = \frac{1 \cdot 2 \cdot 3 \ldots (n-1)}{p(p+1)\ldots(p+n)},$$

and also:

$$B(p,1) = \frac{1}{p}.$$

For any natural numbers m, n we obtain:

$$B(m,n) = \frac{(n-1)!(m-1)!}{(m+n-1)!}.$$

5. Legendre[4] duplication formula:

$$\Gamma(2p) = \frac{2^{2p-1}}{\sqrt{\pi}}\Gamma(p)\Gamma\left(p+\frac{1}{2}\right).$$

Proof Substituting $p = q$ in $B(p,q)$:

$$B(p,q) = \frac{\Gamma(p)\Gamma(q)}{\gamma(p+q)} = \int_0^1 u^{p-1}(1-u)^{q-1}\,du,$$

and replacing $u = \dfrac{1+x}{2}$, we have:

$$\frac{\Gamma^2(p)}{\Gamma(2p)} = \frac{1}{2}\int_{-1}^{1}\left(\frac{1+x}{2}\right)^{p-1}\left(\frac{1-x}{2}\right)^{p-1} = \frac{1}{2^{2p-1}}\int_{-1}^{1}(1-x^2)^{p-1}\,dx,$$

resulting

$$2^{2p-1}\Gamma^2(p) = 2\Gamma(2p)\int_0^1 (1-x^2)^{p-1}\,dx.$$

But $B\left(\frac{1}{2}, p\right) = 2\int_0^1 (1-x^2)^{p-1}dx, \quad \Gamma\left(\frac{1}{2}\right) = \sqrt{\pi}$, resulting finally:

$$2^{2p-1}\Gamma^2(p) = \Gamma(2p)B\left(\frac{1}{2},\ p\right) = \Gamma(2p)\frac{\Gamma\left(\frac{1}{2}\right)\Gamma(p)}{\Gamma\left(\frac{1}{2}+p\right)}.$$

6. Triplication formula:

$$\Gamma(3p) = \frac{3^{3p-\frac{1}{2}}}{\sqrt{\pi}}\Gamma(p)\Gamma\left(p+\frac{1}{2}\right)\Gamma\left(p+\frac{2}{3}\right).$$

[4] A.M. Legendre (1752–1833).

7. Gauss multiplication formula:

$$\Gamma(p)\Gamma\left(p+\frac{1}{k}\right)\ldots\Gamma\left(p+\frac{k-1}{k}\right)=(2k)^{(k-1)/2}k^{-kp+1}\Gamma(kp).$$

1.2 Integral Functions

In this section we introduce the error, imaginary error, complementary error, and exponential integral functions, denoted as erf(·), erfi(·), erfc(·), and $Ei(\cdot)$, respectively.

Function Error function (erf) is defined as:

$$\mathrm{erf}(az)=\frac{2a}{\sqrt{\pi}}\int_0^z e^{-a^2z^2}\,dz.$$

Function Imaginary error function (erfi) is defined as:

$$\mathrm{erfi}(az)=\frac{2a}{\sqrt{\pi}}\int_0^z e^{a^2z^2}dz=-i\ \mathrm{erf}(iaz)$$

Function Complementary error function (erfc) is defined as:

$$\mathrm{erfc}(az)=\frac{2a}{\sqrt{\pi}}\int_z^\infty e^{-a^2z^2}dz$$

Function Exponential integral function (Ei) is defined as:

$$Ei(az)=\int_{-\infty}^z\frac{e^{az}}{z}dz.$$

1.3 Mittag-Leffler Function

In this section we introduce the one- and two-parameter Mittag-Leffler functions, denoted as $E_\alpha(\cdot)$ and $E_{\alpha,\beta}(\cdot)$, respectively.

The one-parameter Mittag-Leffler[5] function, One parameter Mittag-Leffler function (E_α), is defined as:

[5]M.G. Mittag-Leffler (1846–1927).

$$E_\alpha(z) = \sum_{k=0}^{\infty} \frac{z^k}{\Gamma(\alpha k + 1)} \qquad \mathrm{Re}(\alpha) > 0.$$

The two-parameter Mittag-Leffler function, Two parameter Mittag-Leffler function ($E_{\alpha,\beta}$), is defined as:

$$E_{\alpha,\beta}(z) = \sum_{k=0}^{\infty} \frac{z^k}{\Gamma(\alpha k + \beta)} \qquad \mathrm{Re}(\alpha), \mathrm{Re}(\beta) > 0, \beta \in \mathbb{C}.$$

For particular values of α and β it results:

$$E_{0,1}(z) = \sum_{k=0}^{\infty} \frac{z^k}{\Gamma(1)} = \sum_{k=0}^{\infty} z^k, \tag{1.4}$$

$$E_{1,1}(z) = \sum_{k=0}^{\infty} \frac{z^k}{\Gamma(k+1)} = \sum_{k=0}^{\infty} \frac{z^k}{k!} = e^z, \tag{1.5}$$

$$E_{1,2}(z) = \sum_{k=0}^{\infty} \frac{z^k}{\Gamma(k+2)} = \sum_{k=0}^{\infty} \frac{z^k}{(k+1)!} = \frac{1}{z} \sum_{k=0}^{\infty} \frac{z^{k+1}}{(k+1)!} = \frac{e^z - 1}{z}, \tag{1.6}$$

$$E_{1,3}(z) = \sum_{k=0}^{\infty} \frac{z^k}{\Gamma(k+3)} = \sum_{k=0}^{\infty} \frac{z^k}{(k+2)!} = \frac{1}{z^2} \sum_{k=0}^{\infty} \frac{z^{k+2}}{(k+2)!} = \frac{e^z - 1 - z}{z^2}, \tag{1.7}$$

$$E_{1,0}(z) = \sum_{k=0}^{\infty} \frac{z^k}{\Gamma(k)} = z e^z, \tag{1.8}$$

$$E_{1,\frac{3}{2}}(at) = \frac{e^{at}}{\sqrt{at}} \mathrm{erf}(\sqrt{at}), \tag{1.9}$$

$$E_{1,\frac{1}{2}}(at) = \frac{1}{\sqrt{\pi}} + e^{at} \sqrt{at}\, \mathrm{erf}(\sqrt{at}), \tag{1.10}$$

$$E_{1,\frac{5}{2}}(at) = \frac{1}{at} \left[\frac{e^{at}}{\sqrt{at}} \mathrm{erf}(\sqrt{at}) - \frac{2}{\sqrt{\pi}} \right], \tag{1.11}$$

$$E_{1,-\frac{1}{2}}(at) = -\frac{1}{2\sqrt{\pi}} + (at) \left[\frac{1}{\sqrt{\pi}} + \sqrt{at}\, e^{at}\, \mathrm{erf}(\sqrt{at}) \right], \tag{1.12}$$

$$E_{\alpha,\beta}(t) = \frac{1}{\Gamma(\beta)} + t E_{\alpha,\alpha+\beta}(t), \tag{1.13}$$

$$E_{\alpha,\beta}(t) = \sum_{k=0}^{\infty} \frac{t^k}{\Gamma(\alpha k + \beta)} = \sum_{k=-1}^{\infty} \frac{t^{k+1}}{\Gamma(\alpha(k+1) + \beta)} = \sum_{k=-1}^{\infty} \frac{t^k}{\Gamma(\alpha k + \alpha + \beta)}$$

$$= \frac{1}{\Gamma(\beta)} + t E_{\alpha,\alpha+\beta}(t),$$

$$E_{\alpha,\beta}^{(m)}(z) = \frac{d^m}{dz^m} E_{\alpha,\beta}(z) = \sum_{k=0}^{\infty} \frac{(k+m)!}{k!} \frac{t^k}{\Gamma(\alpha k + \alpha m + \beta)}. \tag{1.14}$$

1.4 Function $E(t, \alpha, a)$

The function E function ($E(t, \alpha, a)$) is defined as:

$$E(t, \alpha, a) = \sum_{k=0}^{\infty} \frac{(at)^k}{\Gamma(k + \alpha + 1)} = t^\alpha E_{1,\alpha+1}(at), \tag{1.15}$$

or in the integral form:

$$E(t, \alpha, a) = \frac{1}{\Gamma(\alpha)} \int_0^t \tau^{\alpha-1} e^{a(t-\tau)} d\tau. \tag{1.16}$$

We prove schematically the equivalence between the forms (1.15) and (1.16). We denote the integral:

$$I = \frac{1}{\Gamma(\alpha)} \int_0^t \tau^{\alpha-1} e^{a(t-\tau)}\, d\tau$$

$$= \frac{1}{\Gamma(\alpha)} \int_0^t \tau^{\alpha-1} \left[\sum_{k=0}^{\infty} \frac{a^k (t-\tau)^k}{k!} \right] d\tau$$

$$= \frac{1}{\Gamma(\alpha)} \sum_{k=0}^{\infty} \left[\frac{a^k t^k}{k!} \left(\int_0^t \tau^{\alpha-1} \left(1 - \frac{\tau}{t}\right)^k d\tau \right) \right].$$

For $u = \frac{\tau}{t}$ we obtain:

$$I_1 = \int_0^t \tau^{\alpha-1} \left(1 - \frac{\tau}{t}\right)^k d\tau = \int_0^1 t^{\alpha-1} u^{\alpha-1} (1-u)^k du$$

$$= t^{\alpha} \int_0^1 u^{\alpha-1}(1-u)^k du = t^{\alpha} B(\alpha, k+1) = \frac{t^{\alpha}\Gamma(\alpha)\Gamma(k+1)}{\Gamma(\alpha+k+1)}.$$

It results:

$$I = t^{\alpha} \sum_{k=0}^{\infty} \frac{(at)^k}{\Gamma(k+\alpha+1)} = t^{\alpha} E_{1,\alpha+1}(at) = E(t, \alpha, a).$$

In (1.16) we take $\alpha = \frac{1}{2}$ and $u^2 = a\tau$. It results after calculations:

$$E_{1,\frac{3}{2}}(at) = \frac{e^{at}}{\sqrt{at}} \text{erf}(a\sqrt{t}).$$

Using the definition we obtain:

$$E_{1,\frac{1}{2}}(at) = \sum_{k=0}^{\infty} \frac{(at)^k}{\Gamma\left(k+\frac{1}{2}\right)}.$$

Replacing k with $k+1$ we have:

$$E_{1,\frac{1}{2}}(at) = \sum_{k=-1}^{\infty} \frac{(at)^{k+1}}{\Gamma\left(k+\frac{1}{2}\right)} = (at)\left[\frac{(at)^{-1}}{\Gamma\left(\frac{1}{2}\right)} + \sum_{k=0}^{\infty} \frac{(at)^k}{\Gamma\left(k+\frac{3}{2}\right)}\right]$$

$$= \frac{1}{\sqrt{\pi}} + at E_{1,\frac{3}{2}}(at) = \frac{1}{\sqrt{\pi}} + \sqrt{at} e^{at} \text{erf}(\sqrt{at})$$

results finally (1.10).

In order to obtain (1.11), the parameter k will be replaced by $k-1$ in $E_{1,\frac{5}{2}}(at)$.

To obtain (1.12) the parameter k will be replaced by $k+2$ in $E_{1,-\frac{1}{2}}(at)$.

For further details about the Mittag-Leffler function, readers can check [5].

There are known a series of fractional trigonometric functions, useful to solve FDE, studied in detail in [6].

References

1. Abramowitz, M., & Stegun, I. A. (Eds.). (1965). *Handbook of mathematical functions. Dover books on mathematics*. New York: Dover Publications.
2. Brychkov, Y. A. (2008). *Handbook of special functions derivatives, integrals, series and other formulas*. Boca Raton: Chapman and Hall/CRC.
3. Erdélyi, A. (1954). *Tables of integral transforms* (Vols. 1–2). New York: McGraw-Hill Book Company.
4. Erdélyi, A. (Ed.). (1955). *Higher transcendental functions* (Vols. I–III). New York: McGraw-Hill Book Company

5. Gorenflo, R., Kilbas, A. A., Mainardi, F., & Rogosin, S. V. (2014). *Mittag-Leffler functions, related topics and applications. Springer monographs in mathematics*. Heidelberg: Springer.
6. Lorenzo, C. F., & Hartley, T. T. (2017). *The fractional trigonometry: With applications to fractional differential equations and science*. Hoboken: Wiley & Sons, Inc.
7. Nikiforov, A. F., & Ouvarov, V. (1976). *Eléments de la théorie des fonctions spéciales*. Moscow: Mir Publishers.
8. Podlubny, I. (1998). *Fractional differential equations: An introduction to fractional derivatives, fractional differential equations, to methods of their solution and some of their applications. Mathematics in science and engineering*. San Diego: Academic Press.

Chapter 2
Fractional Derivative and Fractional Integral

2.1 Fractional Integral and Derivative

Definition (Fractional Integral of Order α) For every $\alpha > 0$ and a local integrable function $f(t)$, the *right* FI of order α is defined:

$$ {}_aI_t^\alpha f(t) = \frac{1}{\Gamma(\alpha)} \int_a^t (t-u)^{\alpha-1} f(u)du, \qquad -\infty \le a < t < \infty. \tag{2.1}$$

Alternatively, it can be defined also the *left* FI as:

$$ {}_tI_b^\alpha f(t) = \frac{1}{\Gamma(\alpha)} \int_t^b (u-t)^{\alpha-1} f(u)du, \qquad -\infty < t < b \le \infty. \tag{2.2}$$

For particular values of the a and b parameters, the following cases are known:

- Riemann[1]: $a = 0, \quad b = +\infty$
- Liouville[2]: $a = -\infty, \quad b = 0.$

Definition (Fractional Derivative of Order α) For every α, and $n = \lceil\alpha\rceil$ the Riemann–Liouville derivative of order α can be defined as:

$$ {}_aD_t^\alpha f(t) = \frac{1}{\Gamma(n-\alpha)} \left(\frac{d}{dx}\right)^n \int_a^t (t-u)^{n-\alpha-1} f(u)\, du. \tag{2.3}$$

[1]G.F.B. Riemann (1826–1866).

[2]J. Liouville (1809–1882).

C. Milici et al., *Introduction to Fractional Differential Equations*, Nonlinear Systems and Complexity 25, https://doi.org/10.1007/978-3-030-00895-6_2

Theorem 1 *The following integration rules are valid:*

$$\int_a^b \phi(x)_a I_x^\alpha \psi(x)dx = \int_a^b \psi(x)_x I_b^\alpha \phi(x)\,dx, \tag{2.4}$$

$$\int_a^b f(x)_a D_x^\alpha g(x)dx = \int_a^b g(x)_x D_b^\alpha f(x)\,dx. \tag{2.5}$$

Also, it must be noticed that:

$$_aI_x^\alpha\,{}_aD_x^\alpha f(x) = f(x),$$

where $0 < \alpha < 1$.

Proof We use the Dirichlet[3] theorem, written in the form:

$$\int_a^b dx \int_a^x f(x,t)dt = \int_a^b dt \int_t^b f(x,t)\,dx.$$

- For the case (2.4), we introduce the following notation:

$$f(x,t) = \frac{1}{\Gamma(\alpha)} \frac{\phi(x)\psi(t)}{(x-t)^{1-\alpha}},$$

in the Dirichlet theorem. It results:

$$\int_a^b \phi(x)dx \frac{1}{\Gamma(\alpha)} \int_a^x \frac{\psi(t)}{(x-t)^{1-\alpha}} dt = \int_a^b \psi(t)dt \frac{1}{\Gamma(\alpha)} \int_t^b \frac{\phi(t)}{(x-t)^{1-\alpha}}\,dt,$$

$$\int_a^b \phi(x)_a I_x^\alpha \psi(x)dx = \int_a^b \psi(x)_x I_b^\alpha \phi(x)\,dx$$

- For case (2.5), we introduce in (2.4):

$$_xD_b^\alpha f(x) = \phi(x),\quad {}_aD_x^\alpha g(x) = \psi(x),\quad {}_aI_x^\alpha\,{}_aD_x^\alpha f(x) = f(x).$$

Theorem 2 *The following integration and derivation rules are valid:*

(a) $_aI_t^{\alpha+1}[Df(t)] = {}_aI_t^\alpha f(t) - \dfrac{(t-a)^\alpha}{\Gamma(\alpha+1)} f(a),$

---

[3]J.P.G.L. Dirichlet (1805–1859).

(b) $_aI_t^\alpha[_aD_t^\alpha f(t)] = f(t) - \sum_{k=1}^{n} {}_aD_t^{\alpha-k} f(t)\Big|_{t=a} \frac{(t-a)^{\alpha-k}}{\Gamma(\alpha-k+1)},$

(c) $D[_aI_t^\alpha f(t)] = {}_aI_t^\alpha[Df(t)] + \frac{(t-a)^{\alpha-1}}{\Gamma(\alpha)} f(a),$

(d) $_aI_t^\alpha f(t) = {}_aI_t^{\alpha+p}[D^p f(t)] + \sum_{k=0}^{p-1} \frac{D^k f(a)(t-a)^{\alpha+k}}{\Gamma(\alpha+k+1)}$, *where* p *is a positive integer.*

(e) $D^p[_aI_t^\alpha f(t)] = {}_aI_t^\alpha[D^p f(t)] + \sum_{k=0}^{p-1} \frac{D^k f(a)(t-a)^{\alpha+k}}{\Gamma(\alpha+k+1)}$, *where* p *is positive integer.*

Proof

(a) Integrating by parts, it results:

$$_aI_t^{\alpha+1}[Df(t)] = \frac{1}{\Gamma(\alpha+1)} \int_a^t (t-u)^\alpha f'(u)du = \frac{1}{\Gamma(\alpha+1)}[(t-u)^\alpha f(u)]\Big|_a^t$$
$$+\frac{\alpha}{\Gamma(\alpha+1)} \int_a^t (t-u)^{\alpha-1} f(u)\,du,$$
$$_aI_t^{\alpha+1}[Df(t)] = -\frac{(t-a)^\alpha}{\Gamma(\alpha+1)} f(a) + \frac{\alpha}{\alpha\Gamma(\alpha)} \int_a^t (t-u)^{\alpha-1} f(u)\,du.$$

(b) This formula can be verified by induction, using (a), or:

$$I = {}_aI_t^\alpha[_aD_t^\alpha f(t)] = \frac{1}{\Gamma(\alpha)} \int_a^t (t-u)^{\alpha-1}[_aD_u^\alpha f(u)]\,du,$$
$$I = \frac{1}{\Gamma(\alpha+1)} \int_0^t \frac{d}{dt}(t-u)^\alpha[_aD_u^\alpha f(u)]\,du,$$
$$I = -\frac{(t-a)^\alpha}{\Gamma(\alpha+1)} {}_aD_t^\alpha f(t)\Big|_{t=a} + \frac{1}{\Gamma(\alpha+2)} \int_0^t \frac{d}{dt}(t-u)^{\alpha+1}[_aD_u^{\alpha-1} f(u)]\,du.$$
$$I = -\frac{(t-a)^\alpha}{\Gamma(\alpha+1)} {}_aD_t^\alpha f(t)\Big|_{t=a} - \frac{(t-a)^{\alpha+1}}{\Gamma(\alpha+2)} {}_aD_t^{\alpha-1} f(t)\Big|_{t=a}$$
$$+\frac{1}{\Gamma(\alpha+2)} \int_0^t \frac{d}{dt}(t-u)^{\alpha+1}[_aD_u^{\alpha-1} f(u)]\,du,$$
$$\ldots$$
$$I = f(t) - \sum_{k=1}^{n} {}_aD_t^{\alpha-k} f(t)\Big|_{t=a} \frac{(t-a)^{\alpha-k}}{\Gamma(\alpha-k+1)}.$$

(c) In the FI, we have:

$$ {}_aI_t^\alpha f(t) = \frac{1}{\Gamma(\alpha)} \int_a^t (t-u)^{\alpha-1} f(u)\, du, $$

and we make the change of variable: $u = t - x^{1/\alpha}$ (see also [3]). We obtain:

$$ {}_aI_t^\alpha f(t) = \frac{1}{\Gamma(1+\alpha)} \int_a^{(t-a)^\alpha} f(t - x^{1/\alpha})\, dx. $$

Then, for $t > 0$:

$$ D[{}_aI_t^\alpha f(t)] = \frac{1}{\Gamma(1+\alpha)} \left[\alpha (t-a)^{\alpha-1} f(a) + \int_a^{(t-a)^\alpha} \frac{\partial}{\partial t} f(t - x^{1/\alpha})\, dx \right]. $$

Reversing the change of variable $t - x^{1/\alpha} = u$, we obtain:

$$ D[{}_aI_t^\alpha f(t)] = \frac{1}{\Gamma(1+\alpha)} \left[\alpha (t-a)^{\alpha-1} f(a) + \alpha \int_a^t (t-u)^{\alpha-1} \frac{\partial}{\partial t} f(u)\, du \right]. $$

Hence:

$$ D[{}_aI_t^\alpha f(t)] = \frac{(t-a)^{\alpha-1} f(a)}{\Gamma(\alpha)} + {}_aI_t^\alpha [Df(t)]. $$

(d) Replacing α by $\alpha + 1$ and f by Df in (a), we have:

$$ {}_aI_t^{\alpha+1}[D^2 f(t)] = {}_aI_t^{\alpha+1}[Df(t)] - \frac{(t-a)^\alpha f(a)}{\Gamma(\alpha+1)}. $$

Replacing ${}_aI_t^{\alpha+1}[Df(t)]$ with (a), we obtain:

$$ {}_aI_t^{\alpha+1}[D^2 f(t)] = {}_aI_t^{\alpha}[f(t)] - \frac{(t-a)^\alpha f(a)}{\Gamma(\alpha+1)} - \frac{Df(a)\,(t-a)^{\alpha-1}}{\Gamma(\alpha+1)}. $$

(d) Can be established by repeated iterations.
(e) For $t > 0$, we must differentiate (c):

$$ D^2[{}_aI_t^\alpha f(t)] = D[{}_aI_t^\alpha f(t)] + \frac{Df(a)\,(t-a)^{\alpha-2}}{\Gamma(\alpha-1)}. $$

$$ D^2[{}_aI_t^\alpha f(t)] = {}_aI_t^\alpha f(t) + \frac{[Df(a)\,(t-a)^{\alpha-2}]}{\Gamma(\alpha-1)} + \frac{[Df(a)\,(t-a)^{\alpha-1}]}{\Gamma(\alpha)}, $$

and by repeated iterations we obtain (e).

Theorem 3 *The exponents property:*

$$ {}_aI_t^{\alpha}\, {}_aI_t^{\beta}\, f(t) = {}_aI_t^{\alpha+\beta} f(t). $$

For this theorem we recommend also [4].

Proof For $\alpha > 0,\ \beta > 0$, it results:

$$ I = {}_aI_t^{\alpha}\, {}_aI_t^{\beta} f(t) = \frac{1}{\Gamma(\alpha)} \frac{1}{\Gamma(\beta)} \int_a^t (t-u)^{\alpha-1} \int_a^u (u-v)^{\beta-1} f(v)\, du\, dv. $$

If we apply the Dirichlet equality

$$ \int_a^t \int_a^u f(v) du dv = \int_a^t \int_v^t f(v)\, du\, dv, $$

we obtain:

$$ I = \frac{1}{\Gamma(\alpha)} \frac{1}{\Gamma(\beta)} \int_a^t \int_v^t (t-u)^{\alpha-1} (u-v)^{\beta-1} f(v)\, du dv, $$

$$ u = v + z(t-v), $$

$$ du = (t-v)dz, \qquad t-u = (1-z)(t-v), $$

$$ I = \frac{1}{\Gamma(\alpha)} \frac{1}{\Gamma(\beta)} \int_a^t (t-v)^{\alpha+\beta-1} f(v) \int_0^1 (1-z)^{\alpha-1} z^{\beta-1}\, dv\, dz, $$

but:

$$ \int_0^1 (1-z)^{\alpha-1} z^{\beta-1} du = B(\alpha-1, \beta-1) = \frac{\Gamma(\alpha)\Gamma(\beta)}{\Gamma(\alpha+\beta)}. $$

Finally, it results:

$$ I = \frac{1}{\Gamma(\alpha+\beta)} \int_a^t (t-v)^{\alpha+\beta-1} f(v) dv = {}_aI_t^{\alpha+\beta} f(t). $$

Theorem 4

(a) ${}_aD_t^{\alpha}\left[{}_aI_t^{\beta} f(t)\right] = {}_aD_t^{\alpha-\beta} f(t).$

(b) ${}_aI_t^{\alpha}\left[{}_aD_t^{\beta} f(t)\right] = {}_aI_t^{\alpha-\beta} f(t) - \sum_{k=1}^{m} \frac{(t-a)^{\alpha-k}}{\Gamma(\alpha+1-k)} {}_aD_t^{\beta-k} f(t)\Big|_{t=a}.$

where $m = \lceil \beta \rceil + 1.$

(c) $_aD_t^{\alpha}\left[{}_aD_t^{\beta}f(t)\right] = {}_aD_t^{\alpha+\beta}f(t) - \sum_{k=1}^{m} {}_aD_t^{\beta-k}f(t)\Big|_{t=a}\dfrac{(t-a)^{-\alpha-k}}{\Gamma(1-\alpha-k)}.$

Proof

(a)

$$_aD_t^{\alpha}\left[{}_aI_t^{\beta}f(t)\right] = \frac{d^n}{dt^n}\left[{}_aI_t^{n-\alpha}\left[{}_aI_t^{\beta}f(t)\right]\right] = \frac{d^n}{dt^n}\left[{}_aI_t^{n-(\alpha-\beta)}\right] = {}_aD_t^{\alpha-\beta}.$$

(b)

$$I = {}_aI_t^{\alpha}\left[{}_aD_t^{\beta}f(t)\right] = \frac{1}{\Gamma(\alpha)}\int_a^t (t-u)^{\alpha-1}\left[{}_aD_t^{\beta}f(u)\right]du$$

$$I = \frac{1}{\Gamma(\alpha+1)}\int_a^t \frac{d}{dt}(t-u)^{\alpha}\left[{}_aD_t^{\beta}f(u)\right]du$$

$$I = -\frac{(t-a)^{\alpha}}{\Gamma(\alpha+1)}\left[{}_aD_t^{\beta}f(t)\Big|_{t=a}\right] + \frac{1}{\Gamma(\alpha+2)}\int_a^t \frac{d}{dt}(t-u)^{\alpha+1}\left[{}_aD_t^{\beta-1}f(u)\right]du$$

$$\cdots$$

Example Solve the following FDE with initial value:

$$D^{1/2}y(t) = y(t),$$

$$D^{-1/2}y(0) = -2\sqrt{\pi},$$

transforming it in a first order differential equation.

Solution Using the theorem 4 (c), we obtain:

$$D^{1/2}\left[D^{1/2}y(t)\right] = y'(t) - D^{1/2-1}y(0)\frac{t^{-1/2-1}}{\Gamma(1-1/2-1)} = D^{1/2}y(t) = y(t),$$

$$y'(t) - t^{-3/2} = y(t).$$

Theorem 5 *Linearity property:*

$$_aI_t^{\alpha}[C_1 f(t) + C_2 g(t)] = C_1\, {}_aI_t^{\alpha} f(t) + C_2\, {}_aI_t^{\alpha} g(t),$$

where: C_1 and C_2 are constants and $f(t)$ and $g(t)$ are two arbitrary functions.

Proof

$$\begin{aligned}
{}_aD_t^{\alpha}[C_1\, f(t) + C_2\, g(t)] &= \frac{1}{\Gamma(\alpha)} \int_a^t (t-y)^{\alpha-1}\, [C_1\, f(y) + C_2\, g(y)]\, dy \\
&= C_1 \frac{1}{\Gamma(\alpha)} \int_a^t (t-y)^{\alpha-1} f(y) dy \\
&\quad + C_2 \frac{1}{\Gamma(\alpha)} \int_a^t (t-y)^{\alpha-1} g(y) dy \\
&= C_1\, {}_aI_t^{\alpha} f(t) + C_2\, {}_aI_t^{\alpha} g(t).
\end{aligned}$$

Theorem 6 *If the function $f(t)$ possess continuous derivative, then for $\alpha > 0$, $n = \lceil \alpha \rceil + 1$:*

$$ {}_aI_t^{\alpha} f(t) = \sum_{k=0}^{n-1} \frac{f^{(k)}(t-a)^{k-\alpha}}{\Gamma(k+1-\alpha)} + \frac{1}{\Gamma(n-\alpha)} \int_a^t (t-y)^{n-\alpha-1} f^{(n)}(y)\, dy. $$

Proof In the Cauchy[4] formula:

$$ {}_aI_t^{\alpha} f(t) = \frac{1}{\Gamma(\alpha)} \int_a^t (t-u)^{\alpha-1} f(u) du, $$

it will be applied successively the integration by parts formula (see [5]).

Theorem 7 *We denote ${}_0I_t^{\alpha}$ with I^{α}, for $p \in \mathbb{N}$, $\alpha > 0$. It can be proved that:*

(a) $I^{\alpha}[t^p f(t)] = \sum_{k=0}^{p} \binom{-\alpha}{k} \frac{d^k}{dt^k} t^p\; I^{\alpha+k} f(t),$

(b) $D^{\alpha}[t^p f(t)] = \sum_{k=0}^{p} \binom{\alpha}{k} \frac{d^k}{dt^k} t^p\; D^{\alpha-k} f(t).$

Proof

(a) $I = I^{\alpha}[t^p f(t)] = \dfrac{1}{\Gamma(\alpha)} \displaystyle\int_0^t (t-u)^{\alpha-1} u^p\, f(u)\, du,$

$$ u^p = (t-(t-u))^p = \sum_{k=0}^{p} \frac{(-1)^k p! t^{p-k}}{k!(p-k)!} (t-u)^k, $$

[4] A.L. Cauchy (1789–1857).

$$I=\sum_{k=0}^{p}(-1)^k\frac{p!}{(p-k)!}t^{p-k}\frac{1}{\Gamma(\alpha)}\int_0^t\frac{1}{k!}(t-u)^{\alpha+k-1}f(u)\,du,$$

$$I=\sum_{k=0}^{p}(-1)^k\frac{(\alpha+k-1)\dots\alpha}{k!}\frac{d^k}{dt^k}t^pI^{\alpha+k}f(t),$$

$$I=\sum_{k=0}^{p}(-1)^k\binom{\alpha}{k}\frac{d^k}{dt^k}t^pI^{\alpha+k}f(t).$$

(b) $I=D^\alpha[t^pf(t)]=\dfrac{d^n}{dt^n}I^{n-\alpha}[t^pf(t)]$. We obtain:

$$I=\sum_{k=0}^{p}\binom{-n+\alpha}{k}\sum_{j=0}^{n}\binom{n}{j}\frac{d^{k+j}}{dt^{k=j}}t^pD^{\alpha-k-j}f(t)$$

$$I=\sum_{i=0}^{p}\frac{d^i}{dt^i}t^pD^{n-i}f(t)\sum_{j=0}^{i}\binom{-n+\alpha}{i-j}\binom{n}{j},$$

but

$$\sum_{j=0}^{i}\binom{-n+\alpha}{i-j}\binom{n}{j}=\binom{\alpha}{i},$$

thus:

$$D^\alpha[t^pf(t)]=\sum_{k=0}^{p}\binom{\alpha}{k}\frac{d^k}{dt^k}t^pD^{\alpha-k}f(t).$$

Definition of Caputo Fractional Derivative

Let $\alpha>0,\quad n=\lceil\alpha\rceil$. The Caputo[5] derivative operator of order α is defined as [1, 2]:

$$^C_aD^\alpha_t f(t)=\frac{1}{\Gamma(n-\alpha)}\int_a^t(t-u)^{n-\alpha-1}\left(\frac{d}{du}\right)^nf(u)\,du. \tag{2.6}$$

For $a=0$, we introduce the notation:

$$^CD^\alpha_t f(t)=D^\alpha f(t).$$

[5]M. Caputo (1967–).

Theorem 8 *For $t > 0$, $\alpha \in \mathbb{R}$, $n-1 < \alpha < n$, $n \in \mathbb{N}$, and a function $f(t)$ which obey the conditions of Taylor[6] theorem, the following representation is valid:*

$$ {}_aD_t^{\alpha} f(t) = {}_a^CD_t^{\alpha} f(t) + \sum_{k=0}^{n-1} \frac{f^{(k)}(a)}{\Gamma(1+k-\alpha)}(t-a)^{k-\alpha}. $$

Proof In order to simplify our presentation, we consider $a = 0$.

Because $f(t)$ can be expanded in Taylor series we can write

$$ f(t) = \sum_{k=0}^{n-1} \frac{t^k}{\Gamma(k+1)} f^{(k)}(0) + R_{n-1}, $$

where:

$$ R_{n-1} = \int_0^t \frac{f^{(n)}(y)(t-y)^{n-1}}{(n-1)!} dy = \frac{1}{\Gamma(n)} \int_0^t f^{(n)}(y)(t-y)^{n-1} dy = I^n f^{(n)}. $$

If we apply the operator D^{α} we obtain successively:

$$ D^{\alpha} f(t) = D^{\alpha} \left[\sum_{k=0}^{n-1} \frac{t^k}{\Gamma(k+1)} f^{(k)}(0) + R_{n-1} \right] = \sum_{k=0}^{n-1} \frac{D^{\alpha} t^k}{\Gamma(k+1)} f^{(k)}(0) + D^{\alpha} R_{n-1}, $$

$$ D^{\alpha} f(t) = \sum_{k=0}^{n-1} \frac{t^{k-\alpha}}{\Gamma(k-\alpha+1)} f^{(k)}(0) + I^{n-\alpha} f^n(t), $$

$$ D^{\alpha} f(t) = \sum_{k=0}^{n-1} \frac{t^{k-\alpha}}{\Gamma(k-\alpha+1)} f^{(k)}(0) + D^{\alpha} f(t). $$

The Caputo Fractional Derivative in the Origin

For a function $f(t)$, for which $f(t) = 0$, if $t < 0$, it can be defined:

$$ {}_0^CD_t^{\alpha} f(t) = \frac{1}{\Gamma(n-\alpha)} \int_0^t (t-u)^{n-\alpha-1} f^{(n)}(u)\, du, $$

where $\lceil \alpha \rceil = n$.

Observation

If C is a constant, then:

$$ {}_0^CD_t^{\alpha} C = 0, $$

[6]B. Taylor (1685–1731).

and the Riemann–Liouville FD of C is:

$${}_0D_t^{\alpha} C = \frac{C\,x^{-\alpha}}{\Gamma(1-\alpha)}, \alpha = 1, 2, \ldots.$$

In what follows we note the Caputo derivative in the origin, simply, using the notation $D^{\alpha} f(x)$.

Theorem 9 *If $n - 1 < \alpha < n$, where $n \in \mathbb{N}$, and $\alpha \in \mathbb{R}$, then:*

$$\lim_{\alpha \to n} D^{\alpha} f(t) = f^{(n)}(t),$$

$$\lim_{\alpha \to n-1} D^{\alpha} f(t) = f^{(n-1)}(t) - f^{(n-1)}(0).$$

Proof In the formula

$$D^{\alpha} f(t) = \frac{1}{\Gamma(n-\alpha)} \int_0^t \frac{f^{(n)}(y)dy}{(t-y)^{\alpha+1-n}},$$

we will use the integration by parts, obtaining:

$$\int_0^t u(y)v'(y)\,dy = u(y)\,v(y)\Big|_0^t - \int_0^t u'(y)\,v(y)\,dy,$$

$$u(y) = f^{(n)}(y), \qquad v'(y) = (t-y)^{n-\alpha-1},$$

$$u'(y) = f^{(n+1)}(y), \qquad v(y) = -(t-y)^{n-\alpha}.$$

It results:

$$D^{\alpha} f(t) = \frac{1}{\Gamma(n-\alpha)} \left[-f^{(n)}(y) \frac{(t-y)^{n-\alpha}}{n-\alpha} \Big|_0^t \right.$$
$$\left. + \frac{1}{n-\alpha} \int_0^t (t-y)^{n-\alpha} f^{(n+1)}(y)dy \right].$$

Using the property of Γ function

$$\Gamma(n-\alpha+1) = (n-\alpha)\Gamma(n-\alpha),$$

it results:

$$D^{\alpha} f(t) = \frac{1}{\Gamma(n-\alpha+1)} \left[f^{(n)}(0) + \int_0^t f^{(n+1)}(y)(t-y)^{n-\alpha}\,dy \right],$$

$$\lim_{\alpha \to n} D^{\alpha} f(t) = \left[f^{(n)}(0) + \int_0^t f^{(n+1)}(y) dy \right] = f^{(n)}(0) + f^{(n)}(y) \Big|_0^t = f^{(n)}(t),$$

$$\lim_{\alpha \to n-1} D^{\alpha} f(t) = \left[f^{(n)}(0) + \int_0^t f^{(n+1)}(y)(t-y) dy \right]$$

$$= f^{(n)}(0)t + (t-y) f^{(n)}(y) \Big|_0^t = f^{(n-1)}(t) - f^{(n-1)}(0).$$

Example 1 Let us calculate the FD for $\alpha > 0$, $n-1 < \alpha < n$, $\beta > n-1$ of the function $f(t) = t^{\beta}$ using the definitions, for the case:

1. Riemann–Liouville.
2. Caputo in the origin, using the definition.

Solution

1. For the Riemann–Liouville derivative, we can write:

$$I = D^{\alpha} t^{\beta} = \frac{1}{\Gamma(n-\alpha)} \frac{d^n}{dt^n} \int_0^t u^{\beta} (t-u)^{n-\alpha-1} \, du,$$

and we take:

$$u = v\,t, \qquad du = t\,dv.$$

It follows:

$$I = \frac{1}{\Gamma(n-\alpha)} \frac{d^n}{dt^n} \int_0^t (vt)^{\beta} [(1-v)t]^{n-\alpha-1} t \, dv$$

$$= \frac{1}{\Gamma(n-\alpha)} \frac{d^n}{dt^n} \int_0^t (1-v)^{n-\alpha-1} v^{\beta} t^{n-\alpha+\beta} \, dv,$$

$$I = \frac{1}{\Gamma(n-\alpha)} \int_0^t (1-v)^{n-\alpha-1} v^{\beta} \frac{d^n}{dt^n} t^{n-\alpha+\beta} \, dv,$$

but

$$\frac{d^n}{dt^n} t^{\lambda} = \frac{\Gamma(\lambda+1)}{\Gamma(\lambda-n+1)} t^{\lambda-n},$$

$$B(p,q) = \int_0^1 v^{p-1} (1-v)^{q-1} \, dv,$$

so that it results:

$$I = \frac{1}{\Gamma(n-\alpha)} \frac{\Gamma(n-\alpha+\beta+1)}{\Gamma(-\alpha+\beta+1)} t^{-\alpha+\beta} \int_0^1 (1-v)^{n-\alpha-1} v^{\beta}\, dv,$$

$$\int_0^1 (1-v)^{n-\alpha-1} v^{\beta}\, dv = B(n-\alpha, \beta+1) = \frac{\Gamma(n-\alpha)\Gamma(\beta+1)}{\Gamma(n-\alpha+\beta+1)},$$

$$D^{\alpha} t^{\beta} = I = \frac{\Gamma(\beta+1)}{\Gamma(-\alpha+\beta+1)} t^{\beta-\alpha}.$$

2. In this case we apply the definition of the Caputo derivative of t^{β}:

$$I = D^{\alpha} t^{\beta} = \frac{1}{\Gamma(n-\alpha)} \int_0^t \frac{(u^{\beta})^{(n)}}{(t-u)^{\alpha+1-\beta}}\, du,$$

$$I = \frac{1}{\Gamma(n-\alpha)} \int_0^t \frac{\Gamma(\beta+1)}{\Gamma(\beta-n+1)} u^{\beta-n} (t-u)^{n-\alpha-1}\, du.$$

We use the change of variable $u = vt$, resulting after calculations:

$$du = t\, dv,$$

$$I = \frac{\Gamma(\beta+1)}{\Gamma(n-\alpha)\Gamma(\beta-n+1)} \int_0^1 (uv)^{\beta-n} [(t-v)^{n-\alpha-1}] t\, dv.$$

Finally, we obtain:

$$I = \frac{\Gamma(\beta+1)}{\Gamma(n-\alpha)\Gamma(\beta-n+1)} B(\beta-n+1, n-\alpha) = \frac{\Gamma(\beta+1)}{\Gamma(\beta-\alpha+1)} t^{\beta-\alpha}.$$

Example 2 Find the Riemann–Liouville FI and FD of

$$f(t) = (t-a)^{\beta}.$$

Solution For the FI we apply the Riemann–Liouville definition:

$$I = {}_a I_t^{\alpha} f(t) = \frac{1}{\Gamma(\alpha)} \int_a^t (t-u)^{\alpha-1} (u-a)^{\beta} du.$$

The following change of variable

$$\frac{u-a}{t-a} = v,$$

$$du = (t-a) dv,$$

allows to calculate:

$$I = \frac{(t-a)^{\alpha+\beta}}{\Gamma(\alpha)} \int_0^1 (1-v)^{\alpha-1} v^{\beta} dv = \frac{(t-a)^{\alpha+\beta}}{\Gamma(\alpha)} B(\alpha, \beta+1),$$

$$I = \frac{\Gamma(\beta+1)}{\Gamma(\alpha+\beta+1)} (t-a)^{\alpha+\beta}.$$

For the FD we apply the Riemann–Liouville definition:

$$Df = {}_aD_t^{\alpha}(t-a)^{\beta} = \frac{d^n}{dt^n} {}_aI^{n-\alpha}(t-a)^{\beta},$$

and finally:

$$Df = \frac{\Gamma(\beta+1)}{\Gamma(\beta+n-\alpha+1)} \frac{d^n}{dt^n}(t-a)^{\beta+n-\alpha} = \frac{\Gamma(\beta+1)}{\Gamma(\beta-\alpha+1)}(t-a)^{\beta-\alpha}.$$

Theorem 10 *If $n-1 < \alpha < n$ and if $f(t)$ satisfy the conditions of the Taylor theorem, then:*

$$D^{\alpha} f(t) = \sum_{k=0}^{\infty} \frac{f^{(k)}(0)}{\Gamma(k-\alpha+1)} t^{k-\alpha}.$$

Proof Because $f(t)$ satisfy the conditions of the Taylor theorem, we can apply the Taylor expansion:

$$f(t) = \sum_{k=0}^{\infty} \frac{f^{(k)}(0)}{k!} t^k = \sum_{k=0}^{\infty} \frac{f^{(k)}(0)}{\Gamma(k+1)} t^k.$$

The FD will be

$$D^{\alpha} f(t) = \sum_{k=0}^{\infty} \frac{f^{(k)}(0)}{\Gamma(k+1)} D^{\alpha} t^k = \sum_{k=0}^{\infty} \frac{f^{(k)}(0)}{\Gamma(k+1)} \frac{\Gamma(k+1)}{\Gamma(k-\alpha+1)} t^{k-\alpha},$$

and finally:

$$D^{\alpha} f(t) = \sum_{k=0}^{\infty} \frac{f^{(k)}(0)}{\Gamma(k-\alpha+1)} t^{k-\alpha}.$$

Theorem 11 *The integration and derivation rules are valid:*

$$I^{\alpha}[t^r f(t)] = \sum_{k=0}^{r} \binom{-\alpha}{k} \frac{d^k}{dt^k} t^r I^{\alpha+k} f(t), \quad r \in \mathbb{N}, \quad \alpha > 0$$

$$D^{\alpha}[t^r f(t)] = \sum_{k-0}^{r} \binom{\alpha}{k} \frac{d^k}{dt^k} t^r D^{\alpha-k} f(t), \quad r \in \mathbb{N}, \quad \alpha \in \mathbb{R}.$$

Proof Using the definition

$$I^{\alpha}[t^r f(t)] = \frac{1}{\Gamma(\alpha)} \int_0^t (t-\tau)^{\alpha-1} \tau^r f(\tau) d\tau$$

and taking

$$\tau^r = [t-(t-\tau)]^r = \sum_{k=0}^{r} \frac{(-1)^k r!}{k!(r-k)!} t^{r-k} (t-r)^k,$$

we obtain:

$$I^{\alpha}[t^r f(t)] = \sum_{k=0}^{r} \frac{(-1)^k r!}{k!(r-k)!} t^{r-k} \frac{1}{\Gamma(\alpha)} \int_0^t \frac{1}{k!} (t-\tau)^{\alpha+k-1} f(\tau) d\tau.$$

But:

$$I^{\alpha+k} f(t) = \frac{1}{\Gamma(\alpha+k)} \int_0^t (t-\tau)^{\alpha+k-1} f(\tau) d\tau,$$

$$\Gamma(\alpha+k) = (\alpha+k-1)\Gamma(\alpha+k-1) = \ldots = (\alpha+k-1)\ldots\alpha\Gamma(\alpha).$$

Finally, we obtain:

$$\begin{aligned} I^{\alpha}[t^r f(t)] &= \sum_{k=0}^{r} (-1)^k \frac{(\alpha+k-1)\ldots\alpha}{k!} \frac{d^k}{dt^k} t^r I^{\alpha+k} f(t) \\ &= \sum_{k=0}^{r} \binom{-\alpha}{k} \frac{d^k}{dt^k} t^r I^{\alpha+k} f(t). \end{aligned}$$

Observation

For $0 < \alpha < 1, \quad f(0) = 0$, we have:

$$I^{\alpha} f(t) = \frac{1}{\Gamma(\alpha)} \int_0^t (t-u)^{\alpha-1} f(u)\, du,$$

$$I^{\alpha} f(t) = \frac{\alpha}{\alpha\Gamma(\alpha)} \int_0^t (t-u)^{\alpha-1} f(u)\, du = \frac{1}{\Gamma(\alpha+1)} \int_0^t f(u)\, (du)^{\alpha}.$$

References

1. Caputo, M. (1999). *Lessons on seismology and rheological tectonics*. Technical report, Universitá degli Studi La Sapienza, Rome.
2. Ishteva, M., Scherer, R., & Boyadjiev, L. (2003). *On the Caputo operator of fractional calculus and C-Laguerre functions*. Technical report, Bulgarian Ministry of Education and Science, Grant MM 1305/2003.
3. Miller, K. S., & Ross, B. (1993). *An introduction to the fractional calculus and fractional differential equations*. New York: John Wiley & Sons.
4. Valério, D., & Costa, J. S. (2013). *An introduction to fractional control*. London: CRC Press.
5. Podlubny, I. (1998). *Fractional differential equations: An introduction to fractional derivatives, fractional differential equations, to methods of their solution and some of their applications. Mathematics in science and engineering*. San Diego: Academic Press.

Chapter 3
The Laplace Transform

A function $f(t)$ is called original function if [8, 9]:

1. $f(t) \equiv 0 \quad$ for $t < 0$,
2. $|f(t)| < Me^{s_0 t}$ for $t > 0$ with $M > 0$, $s_0 \in \mathbb{R}$.
3. For every closed interval $[a, b]$, the function satisfies the Dirichlet conditions:

   (a) is bounded,
   (b) or is continuous, or has a finite number of discontinuities of first kind,
   (c) has a finite number of extremes.

We consider the complex variable $s = \alpha + i\beta$, where $\text{Re}(s) = \alpha \geq s_1 \geq s_0$. Then

$$F(s) = \int_0^\infty e^{-st} f(t)\, dt, \tag{3.1}$$

is called the Laplace[1] integral, or Laplace transform (LT), or *image* of the original function $f(t)$. In the follow-up we denote by $L[f(t)] = F(s)$ or simply by Laplace transform (L) the Laplace transform. In Table 3.1 the LT of some elementary usual functions are listed.

The corresponding inverse Laplace transform is [1, 4]:

$$f(t) = \frac{1}{2\pi i} \lim_{t\to\infty} \int_{\gamma - it}^{\gamma + it} F(s)e^{st} dt = L^{-1}[F(s)], \tag{3.2}$$

where $i = \sqrt{-1}$ and $\gamma \in \mathbb{R}$, so that the contour path of integration is contained in the convergence region.

[1] Pierre Laplace (1749–1827).

C. Milici et al., *Introduction to Fractional Differential Equations*, Nonlinear Systems and Complexity 25, https://doi.org/10.1007/978-3-030-00895-6_3

Table 3.1 Images of basic elementary functions

Number	Original	Image	Number	Original	Image
1	1	$\frac{1}{s}$	7	$e^{\alpha t}\cos\beta t$	$\frac{s-\alpha}{(s-\alpha)^2+\beta^2}$
2	$\frac{t^n}{n!}$	$\frac{1}{s^{n+1}}$	8	$e^{\alpha t}\sin\beta t$	$\frac{\beta}{(s-\alpha)^2+\beta^2}$
3	$e^{\alpha t}$	$\frac{1}{s-\alpha}$	9	$\frac{t^n}{n!}e^{\alpha t}$	$\frac{1}{(s-\alpha)^{n+1}}$
4	$\cos\beta t$	$\frac{s}{s^2+\beta^2}$	10	$t\cos\beta t$	$\frac{s^2-\beta^2}{(s^2+\beta^2)^2}$
5	$\sin\beta t$	$\frac{\beta}{s^2+\beta^2}$	11	$t\sin\beta t$	$\frac{2s\beta}{(s^2+\beta^2)^2}$
6	$\cosh(\beta t)$	$\frac{s}{s^2-\beta^2}$	12	$\sinh(\beta t)$	$\frac{\beta}{s^2-\beta^2}$

3.1 Calculus of the Images

Example 1 Establish the image of $f(t) = t^\lambda$:

$$F(s) = \int_0^\infty e^{-ts} t^\lambda \, dt.$$

We introduce the change of variable $x = ts$. We have also $dx = s\,dt$. It results:

$$F(s) = \int_0^\infty e^{-x} \frac{x^\lambda}{s^\lambda} \frac{dx}{s},$$

$$F(s) = \frac{1}{s^{\lambda+1}} \int_0^\infty e^{-x} x^\lambda dx = \frac{\Gamma(\lambda+1)}{s^{\lambda+1}}.$$

The direct and inverse LT are:

$$L(t^\lambda) = \frac{\Gamma(\lambda+1)}{s^{\lambda+1}}, \qquad L^{-1}\left(\frac{1}{s^{\lambda+1}}\right) = \frac{t^\lambda}{\Gamma(\lambda+1)}.$$

Example 2 Find the image of: $f(t) = \sin^2(t)$.

Using the identity $\sin^2 t = \dfrac{1-\cos(2t)}{2}$, it results:

$$F(s) = \frac{1}{2s} - \frac{2s}{2(s^2+4)} = \frac{2}{s(s^2+4)}.$$

Example 3 Find the image of $f(t) = \frac{1}{2}te^{bt} + \frac{1}{2}te^{-bt}$.

$$F(s) = \frac{1}{2(s-b)^2} + \frac{1}{2(s+b)^2} = \frac{s^2+b^2}{(s^2-b^2)^2}.$$

3.2 Calculus of the Original Function

3.2.1 Calculus of Original Using Residues

If we denote by $L[f(t)] = F(s)$, then if we consider all residues of the function $F(s)e^{st}$, denoted by $r_1, r_2, \ldots r_n$ we can use the theorem:

$$f(t) = r_1 + r_2 + \ldots + r_n.$$

The residues, denoted by Residues (Res), can be calculated using the following procedure (theorem):

If a is a simple pole of the function, then:

$$\operatorname*{Res}_{a}[e^{st}F(s)] = \lim_{s\to a}[(s-a)e^{st}F(s)].$$

If a is a simple pole of order n of the function, then:

$$\operatorname*{Res}_{a}[e^{st}F(s)] = \frac{1}{(n-1)!}\lim_{s\to a}[(s-a)^n e^{st}F(s)]^{(n-1)}.$$

Example 1 Find the original function of the image:

$$F(s) = \frac{1}{(s-3)^2(s+1)}.$$

Solution The residue of the function $F(s)e^{st}$ is

$$\begin{aligned}
r_1 &= \lim_{t\to -1}(s+1)\frac{e^{st}}{(s-3)^2(s+1)} = \frac{e^{-t}}{16},\\
r_2 &= \lim_{t\to 3}\frac{d}{ds}\left(\frac{e^{st}}{s+1}\right)\\
&= \frac{te^{3t}-e^{3t}}{16},
\end{aligned}$$

resulting:

$$f(t) = r_1 + r_2 = \frac{e^{-t}}{16} + \frac{te^{3t}-e^{3t}}{16}.$$

Example 2 Find the original function of the following image:

$$F(s) = \frac{s^2}{s^4-1}.$$

Solution The function F has singularities: $1,\ -1,\ -i,\ i$. The residues will be:

$$r_1 = \operatorname*{Res}_{1} F(s)e^{st} = \lim_{s\to 1}(s-1)\frac{s^2 e^{st}}{(s-1)(s+1)(s^2+1)} = \frac{e^t}{4},$$

$$r_2 = \operatorname*{Res}_{-1} F(s)e^{st} = \lim_{s\to -1}(s+1)\frac{s^2 e^{st}}{(s-1)(s+1)(s^2+1)} = -\frac{e^{-t}}{4},$$

$$r_3 = \operatorname*{Res}_{-i} F(s)e^{st} = \lim_{s\to -i}(s+i)\frac{s^2 e^{st}}{(s^2-1)(s-i)(s+i)} = \frac{-e^{-it}}{4i},$$

$$r_4 = \operatorname*{Res}_{i} F(s)e^{st} = \lim_{s\to i}(s+i)\frac{s^2 e^{st}}{(s^2-1)(s-i)(s+i)} = \frac{e^{it}}{4i}.$$

It results finally

$$f(t) = \frac{1}{2}\left(\frac{e^t - e^{-t}}{2}\right) + \frac{1}{2}\left(\frac{e^{it} - e^{-it}}{2i}\right),$$

or:

$$f(t) = \frac{1}{2}(\sinh t + \sin t).$$

3.2.2 *Calculus of Original with Post's Inversion Formula*

E. Post[2] obtained the formula [7, 10]:

$$f(t) = \lim_{k\to\infty} \frac{(-1)^k}{k!}\left(\frac{k}{t}\right)^{k+1} F^{(k)}\left(\frac{k}{t}\right), \quad t > 0.$$

Example 3 Find the original function of the image:

$$F(s) = \frac{n!}{s^{n+1}} n! s^{-1-n}.$$

Solution With the aid of Post formula we have

$$f(t) = t^n \lim_{k\to\infty} \frac{k^{k+1}(n+k)!}{k! t^{k+1}} \left(\frac{k}{t}\right)^{-n-k-1}.$$

[2]E. Post (1897–1954).

Using the Stirling[3] formula:

$$\lim_{k\to\infty} \frac{k!}{\sqrt{2\pi k}} k^k e^{-k} = 1,$$

it results:

$$f(t) = t^n e^{-n} \lim_{k\to\infty} \sqrt{1+\frac{n}{k}} \left(1+\frac{n}{k}\right)^k \left(1+\frac{n}{k}\right)^n = t^n.$$

3.3 The Properties of the Laplace Transform

In this section, we will use the notations $F(s) = L[f(t)]$ and $L(s) = G[g(t)]$. In the follow-up are discussed properties of the LT.

3.3.1 The Property of Linearity

$$L[af(t) + bg(t)] = aF(s) + bG(s), \quad a, b \in \mathbb{R}. \tag{3.3}$$

3.3.2 Similarity Theorem

$$L[f(\alpha t)] = \frac{1}{\alpha} F\left(\frac{s}{\alpha}\right), \qquad \alpha > 0. \tag{3.4}$$

3.3.3 The Differentiation and Integration Theorems

Theorem (Differentiation of an Original) *The LT of the derivative of order k from $f(t)$ gives:*

$$L[f^{(k)}(t)] = s^k F(s) - \left[s^{k-1} f(0) + s^{k-2} f'(0) + \ldots + f^{(k-1)}(0)\right]. \tag{3.5}$$

Proof For $k = 1$, using the definition and integrating by parts, we have:

$$\begin{aligned} L[f'(t)] &= \int_0^\infty e^{-st} f'(t)dt = e^{-st} f(t)\Big|_0^\infty + s \int_0^\infty e^{-st} f(t)\, dt \\ &= -f(0) + sL[f(t)], \end{aligned}$$

---

[3]J. Stirling (1692–1770).

$$L[f'(t)] = sF(s) - f(0),$$

and for $k = 2$, using $L[f''(t)] = L[(f'(t))']$ we obtain:

$$L[f''(t)] = s^2F(s) - sf(0) - f'(0).$$

Using mathematical induction method, we have: $L[f^{(k)}(t)]$.

Example 1 Find the LT for original $f(t) = t^2$.

Solution

$$f'(t) = 2t, \quad f''(t) = 2,$$
$$f(0) = 0, \quad f'(0) = 0, \quad f''(0) = 2,$$
$$L[f''(0)] = s^2F(s) - sf(0) - f'(0),$$
$$L[2] = \frac{2}{s} = s^2F(s) \quad \Rightarrow \quad F(s) = \frac{2}{s^3}.$$

Example 2 Find the LT of following original $f(t) = \cos 2t$.

Solution $f'(t) = -2\sin(2t), \quad f''(t) = -4\cos(2t),$

$$f(0) = 1, \quad f'(0) = 0, \quad f''(0) = -4,$$
$$L[f''(t)] = s^2F(s) - sf'(0) - f(0), \quad \Rightarrow \quad -4F(s) = s^2F(s) - s,$$

$$F(s) = \frac{s}{s^2+4}.$$

Theorem (Integration of an Original) *It can be obtained:*

$$L\left[\int_0^t f(\tau)d\tau\right] = \frac{F(s)}{s},$$

Proof Let be $g(t) = \int_0^t f(\tau)d\tau$. Then:

$$L[g'(t)] = sL[g] - g(0).$$

Also, we have $g'(t) = f(t)$ and $g(0) = 0$. Then:

$$L[g] = \frac{F(s)}{s}.$$

Theorem (Differentiation of a Transform) *We have:*

$$F^{(n)}(s) = L[(-t)^n f(t)].$$

Proof The theorem can be proved by induction. For $n = 1$, we have, successively:

$$F'(s) = L[-tf(t)],$$

$$\frac{dF(s)}{ds} = \frac{d}{ds}\int_0^\infty e^{-st} f(t)dt = -\int_0^\infty e^{-st} tf(t)dt = -L[tf(t)],$$

and, finally:

$$F^{(n)}(s) = \frac{d}{ds}[F^{(n-1)}(s)].$$

3.3.4 Delay Theorem

For a positive number a we have:

$$L[f(t-a)] = e^{-as}F(s)$$

3.3.5 Displacement Theorem

It is valid the formula:

$$L[e^{\lambda t}f(t)] = F(s-\lambda).$$

3.3.6 Multiplication Theorem

The convolution product of two functions $f(t)$ and $g(t)$ is designated by the symbol $*$. We have:

$$(f * g)(t) = \int_0^t f(\tau)g(t-\tau)\,d\tau,$$

$$L[(f * g)(t)] = F(s)G(s), \quad F(s) = L[f(t)], \quad G(s) = L[g(t)].$$

3.3.7 Properties of the Inverse Laplace Transform

The following formula is valid:

The inverse LT is not unique. We have:

$$L^{-1}\left[\frac{s^{-(\alpha-\beta)}}{s^\beta - a}\right] = t^{\alpha-1}E_{\beta,\alpha}(at^\beta),\ \alpha, \beta > 0, s^\alpha > |a|, \tag{3.6}$$

$$L^{-1}\left[\frac{s^{-(\alpha-1)}}{s-a}\right]=t^{\alpha-1}E_{1,\alpha}(at)=E(t,\alpha-1,a), \tag{3.7}$$

$$L^{-1}\left[\frac{s^{-\alpha}}{(s-a)^2}\right]=tE(t,\alpha,a)-\alpha E(t,\alpha+1,a), \tag{3.8}$$

$$L^{-1}\left[\frac{s^{-\alpha}}{(s-a)^3}\right]=\frac{1}{2}t^2E(t,\alpha,a)-\alpha tE(t,\alpha+1,a)+\frac{\alpha(\alpha+1)}{2}E(t,\alpha+2,a), \tag{3.9}$$

$$L^{-1}\left[\frac{1}{(s^\alpha+as^\beta)^{n+1}}\right]=t^{\alpha(n+1)-1}\sum_{k=0}^{\infty}\frac{(-a)^k\binom{n+k}{k}}{\Gamma[k(\alpha-\beta)+(n+1)\alpha]}t^{k(\alpha-\beta)}, \tag{3.10}$$

where $0<\beta\le\alpha$.

$$L^{-1}\left[\frac{s^\gamma}{s^\alpha+as^\beta+b}\right]=t^{\alpha-\gamma-1}\sum_{n=0}^{\infty}\sum_{k=0}^{\infty}\frac{(-b)^n(-a)^k\binom{n+k}{k}}{\Gamma[k(\alpha-\beta)+(n+1)\alpha-\gamma]}t^{k(\alpha-\beta)+n\alpha}, \tag{3.11}$$

where $\beta\le\alpha,\ \ \gamma<\alpha,\ a\in\mathbb{R}$ or: $|a|<s^{\alpha-\beta},\ |b|<|s^\alpha+as^\beta|$.

Proof Proof of the identity (3.6):

$$\begin{aligned}L\left[t^{\alpha-1}E_{\beta,\alpha}(at^\beta)\right]&=\int_0^\infty e^{-st}t^{\alpha-1}E_{\beta,\alpha}(at^\beta)dt\\&=\sum_{k=0}^{\infty}\frac{a^k}{\Gamma(k\beta+\alpha)}\int_0^\infty e^{-st}t^{k\beta+\alpha-1}dt\\&=\sum_{k=0}^{\infty}\frac{a^k}{\Gamma(k\beta+\alpha)}L[t^{k\beta+\alpha-1}]\\&=\sum_{k=0}^{\infty}\frac{a^k}{\Gamma(k\beta+\alpha)}\frac{\Gamma(k\beta+\alpha)}{s^{k\beta+\alpha}}=\frac{1}{s^\alpha}\sum_{k=0}^{\infty}\left(\frac{a}{s^\beta}\right)^k=\frac{s^{-(\alpha-\beta)}}{s^\beta-a}.\end{aligned}$$

Proof of the identity (3.8):

$$\begin{aligned}L[t\,E(t,\alpha,a)-\alpha E(t,\alpha+1,a)]&=-\frac{d}{ds}L[E(t,\alpha,a)]-\alpha L[E(t,\alpha+1,a)]\\&=-\frac{d}{ds}\left[\frac{s^{-\alpha}}{s-a}\right]-\alpha\left[\frac{s^{-(\alpha+1)}}{s-a}\right]=\frac{1}{s^\alpha(s-a)^2}.\end{aligned}$$

Proof of the identity (3.9):

$$L\left[\frac{1}{2}t^2E(t,\alpha,a)-\alpha tE(t,\alpha+1,a)+\frac{1}{2}\alpha(\alpha+1)E(t,\alpha+2,a)\right]$$

$$=\frac{1}{2}\frac{d^2}{ds^2}L\left[E(t,\alpha,a)\right]+\alpha\frac{d}{ds}L\left[E(t,\alpha+1,a)\right]+\frac{\alpha(\alpha+1)}{2}L\left[E(t,\alpha+2,a)\right]$$

$$=\frac{1}{2}\frac{d^2}{ds^2}\left(\frac{s^{-\alpha}}{s-a}\right)+\alpha\frac{d}{ds}\left(\frac{s^{-(\alpha+1)}}{s-a}\right)+\frac{\alpha(\alpha+1)}{2}\left(\frac{s^{-(\alpha+2)}}{s-a}\right)=\frac{1}{s^{\alpha}(s-a)^3}.$$

For the identities (3.10) and (3.11) the reader can use the reference [5].
Proof of the identity (3.10). We will apply the well-known identity [4]:

$$\frac{1}{(1+x)^n}=\sum_{k=0}^{\infty}\binom{n+k}{k}(-x)^k.$$

It follows:

$$\frac{1}{(s^{\alpha}+as^{\beta})^{n+1}}=\frac{1}{(s^{\alpha})^{n+1}}\frac{1}{\left(1+\frac{a}{s^{\alpha\beta}}\right)^{n+1}}$$

$$=\frac{1}{(s^{\alpha})^{n+1}}\sum_{k=0}^{\infty}\binom{n+k}{k}\left(\frac{-a}{s^{\alpha-\beta}}\right)^k.$$

Proof of the identity (3.11):

$$\frac{s^{\gamma}}{s^{\alpha}+as^{\beta}+b}=\frac{s^{\gamma}}{s^{\alpha}+as^{\beta}}\frac{1}{1+\dfrac{b}{s^{\alpha}+as^{\beta}}}=\sum_{n=0}^{\infty}\frac{s^{\gamma}(-b)^n}{(s^{\alpha}+as^{\beta})^{n+1}},$$

and for the case of (3.10) we obtain:

$$\frac{s^{\gamma}}{(s^{\alpha}+s^{\beta}a)^{n+1}}=\frac{s^{\gamma}}{s^{\alpha(n+1)}}\frac{1}{\left(1+\frac{a}{s^{\alpha-\beta}}\right)^{n+1}}$$

$$=\frac{1}{s^{\alpha(n+1)-\gamma}}\sum_{k=0}^{\infty}\binom{n+k}{k}\left(\frac{-a}{s^{\alpha-\beta}}\right)^k=\sum_{k=0}^{\infty}\binom{n+k}{k}\frac{(-a)^k}{s^{\alpha(n+1)+k(\alpha-\beta)-\gamma}},$$

$$\frac{s^{\gamma}}{s^{\alpha}+as^{\beta}+b}=\sum_{n=0}^{\infty}(-b)^n\sum_{k=0}^{\infty}\binom{n+k}{k}\frac{(-a)^k}{s^{\alpha(n+1)+k(\alpha-\beta)-\gamma}},$$

$$L^{-1}\left[\frac{s^{\gamma}}{s^{\alpha}+as^{\beta}+b}\right]=t^{\alpha-\gamma-1}\sum_{n=0}^{\infty}\sum_{k=0}^{\infty}\frac{(-b)^n(-a)^k\binom{n+k}{k}}{\Gamma[k(\alpha-\beta)+(n+1)\alpha-\gamma]}t^{k(\alpha-\beta)+n\alpha}.$$

Lemma *The following identities are valid:*

$$L^{-1}\left[\frac{1}{s^{\alpha}+as+b}\right]=t^{\alpha-1}\sum_{n=0}^{\infty}\sum_{k=0}^{\infty}\frac{(-b)^n(-a)^k\binom{n+k}{k}}{\Gamma[k(\alpha-1)+(n+1)\alpha]}t^{k(\alpha-1)+n\alpha}, \tag{3.12}$$

for $1\leq\alpha$, $0<\alpha$, $a\in\mathbb{R}$, *and* $|a|<s^{\alpha-1}$, $|b|<|s^{\alpha}+as|$, *respectively:*

$$L^{-1}\left[\frac{s^{\alpha-1}}{s^{\alpha}+as+b}\right]=\sum_{n=0}^{\infty}\sum_{k=0}^{\infty}\frac{(-b)^n(-a)^k\binom{n+k}{k}}{\Gamma[k(\alpha-1)+n\alpha+1]}t^{k(\alpha-1)+n\alpha}, \tag{3.13}$$

for $1\leq\alpha$, $a\in\mathbb{R}$ *and for* $|a|<s^{\alpha-1}$, $|b|<|s^{\alpha}+as|$.

Proof Proof of the identity (3.12). In (3.11) we take $\gamma=0$, $\beta=1$.
Proof of the identity (3.13). In (3.11) we take $\gamma=\alpha-1$, and $\beta=1$.

Example 1 Establish the LT of:

$$f(t)=y''(t)-2y'(t)-3y(t); \qquad \text{where: } y(0)=y'(0)=0.$$

Solution

$$F(s)=s^2Y(s)-sy(0)-y'(0)-2[sY(s)-y(0)]-3Y(s),$$

and finally:

$$F(s)=(s^2-2s-3)Y(s).$$

Example 2 Establish the LT of:

$$y=\int_0^t ydt+1.$$

Solution

$$Y(s)=\frac{Y(s)}{s}+\frac{1}{s}\Rightarrow Y(s)=\frac{1}{s-1}.$$

Example 3 Establish the LT of:

$$\int_0^t y(\tau)\sin(t-\tau)d\tau=1-\cos t.$$

Solution

$$Y(s)\frac{1}{s^2+1}=\frac{1}{s}-\frac{s}{s^2+1}=\frac{1}{s(s^2+1)},\qquad \Rightarrow Y(s)=\frac{1}{s}.$$

Example 4 Establish the LT of:

$$\int_0^t y(\tau)e^{t-\tau}d\tau = y(t)-e^t.$$

Solution

$$Y(s)\frac{1}{s-1}=Y(s)-\frac{1}{s-1};\qquad \Rightarrow \qquad Y(s)=\frac{1}{s-2}.$$

3.4 Laplace Transform of the Fractional Integrals and Derivatives

3.4.1 Fractional Integrals

If $\alpha > 0$, the Riemann–Liouville and Caputo FI are the same for both cases:

$$I = I^\alpha f(t) = \frac{1}{\Gamma(\alpha)}\int_0^t (t-y)^{\alpha-1} f(y)dy.$$

Using the LT of the convolution product formula, we have:

$$L[I]=\frac{1}{\Gamma(\alpha)}L[t^{\alpha-1}]L[f(t)]=\frac{F(s)}{s^\alpha}.$$

3.4.2 Fractional Derivatives

– The Riemann–Liouville FD is

$$L\left[D_t^\alpha f(t)\right]=L\left[\frac{1}{\Gamma(n-\alpha)}\left(\frac{dt^n}{d^nt}\right)\int_0^t (t-u)^{n-\alpha-1}f(u)du\right]$$
$$=L\left[\left(\frac{dt^n}{d^nt}\right)I^{n-\alpha}f(t)\right],$$

where we can apply the classical formula:

$$L[f^{(n)}(t)]=s^nF(s)-s^{n-1}f'(0)-\ldots-f^{(n-1)}(0),$$

$$L[D_t^{\alpha} f(t)] = s^n \frac{F(s)}{s^{n-\alpha}} - \sum_{k=0}^{n-1} s^{n-\alpha-1} [D^k I^{n-\alpha} f(t)]_{t=0},$$

$$L[D_t^{\alpha} f(t)] = s^{\alpha} F(s) - \sum_{k=0}^{n-1} s^{n-\alpha-1} [D^k I^{n-\alpha} f(t)]_{t=0}.$$

– The Caputo [2, 3] FD is

$$L[D^{\alpha} f(t)] = L\left[\frac{1}{\Gamma(n-\alpha)} \int_0^t (t-u)^{n-\alpha-1} f^{(n)}(u) du\right] = L[I^{n-\alpha} f^{(n)}(t)],$$

where we can apply the classical formula:

$$L[f^{(n)}(t)] = s^n F(s) - s^{n-1} f'(0) - \ldots - f^{(n-1)}(0),$$

and

$$L[D^{\alpha} f(t)] = s^n \frac{F(s)}{s^{n-\alpha}} - \sum_{k=0}^{n-1} s^{n-\alpha-1} [I^{n-\alpha} f^{(k)}(t)]_{t=0}.$$

$$L[D^{\alpha} f(t)] = s^{\alpha} F(s) - \sum_{k=0}^{n-1} s^{n-\alpha-1} f^{(k)}(0).$$

Exercise 5 For the function $f(t) = t^2$, calculate the Caputo $L[D^{\alpha}]$. It results:

1. $\alpha = \frac{1}{2}$,
2. $\alpha = -\frac{1}{2}$.

Solution

1.

$$L[D^{1/2} t^2] = \frac{1}{s^{1-\frac{1}{2}}} L[2t],$$

$$L[D^{1/2} t^2] = \frac{2}{s^{\frac{5}{2}}},$$

$$D^{1/2} t^2 = L^{-1}\left[\frac{2}{s^{\frac{5}{2}}}\right] = \frac{2t^{\frac{3}{2}}}{\Gamma\left(\frac{5}{2}\right)} = \frac{8t^{\frac{3}{2}}}{3\sqrt{\pi}}.$$

In **MAPLE**, the FD of order $1/2$ of t from t^2 can be evaluated using the command: **fracdiff(t^2,t,1/2)**.

For $0 < \alpha < 1$, $f^{\alpha}(0) = 0$, and

$$F(t) = \int_0^t f(u)(du)^{\alpha} = \alpha \int_0^t (t-u)^{\alpha-1} f(u)\,du$$

we define:

$$L_{\alpha}[f(t)] = F_{\alpha}(s) = \int_0^{\infty} E_{\alpha}(-s^{\alpha}t^{\alpha}) f(t)\,(dt)^{\alpha}.$$

The following formulae can be obtained without difficulty [6]:

1. $L_{\alpha}[t^{\alpha} f(t)] = -D^{\alpha} L_{\alpha}[f(t)]$.
2. $L_{\alpha}[f(at)] = \dfrac{1}{a^{\alpha}} L_{\alpha}[f(t)]$.
3. $L_{\alpha}[f(t-b)] = E_{\alpha}(-s^{\alpha}b^{\alpha}) L_{\alpha}[f(t)]$.
4. $L_{\alpha}\left[\int_0^t f(u)\,(du)^{\alpha}\right] = \dfrac{1}{a^{\alpha}\Gamma(\alpha+1)} L_{\alpha}[f(t)]$.
5. $L_{\alpha}[g^{(\alpha)}(t)] = s^{\alpha} L_{\alpha}[g(t)] - \Gamma(\alpha+1) g(0)$.

For

$$(f(t) * g(t))_{\alpha} = \int_0^t f(t-u) g(u) (du)^{\alpha},$$

we have:

$$L_{\alpha}\left[(f(t) * g(t))_{\alpha}\right] = L_{\alpha}[f(t)]\, L_{\alpha}[g(t)].$$

References

1. Abramowitz, M., & Stegun, I. A. (Eds.). (1965). *Handbook of mathematical functions. Dover books on mathematics*. New York: Dover Publications.
2. Caputo, M. (1999). *Lessons on seismology and rheological tectonics*. Technical report, Universitá degli Studi La Sapienza, Rome.
3. Caputo, M., & Fabrizio, M. (2015). Damage and fatigue described by a fractional derivative model. *Journal of Computational Physics, 293*, 400–408.
4. Debnath, L., & Bhatta, D. (2007). *Integral transforms and their applications*. Boca Raton: Chapman and Hall/CRC.
5. Erdélyi, A. (Ed.). (1955). *Higher transcendental functions* (Vols. I–III). New York: McGraw-Hill Book Company
6. Jumarie, G. (2009). Laplaces transform of fractional order via the Mittag-Leffler function and modified Riemann-Liouville derivative. *Applied Mathematics Letters, 22*, 1659–1664.
7. Post, E. L. (1930). Generalized differentiation. *Transactions of the American Mathematical Society, 32*, 723–781.

8. Poularikas, A. D. (Ed.). (2000). *The transforms and applications handbook* (2nd ed.). CRC Press: Boca Raton.
9. Schiff, J. L. (Ed.). (1999). *The laplace transform, theory and applications*. New York: Springer.
10. Wheeler, N. (1997). *Construction and physical application of fractional calculus*. Technical report, Reed College Physics Department.

Chapter 4
Fractional Differential Equations

4.1 The Existence and Uniqueness Theorem for Initial Value Problems

Definition 1 Let be the fractional differential equation (FDE)

$$(D^{\alpha}_{a_+} y)(t) = f[t, y(t)], \quad \alpha > 0, \quad t > a,$$

with the conditions:

$$(D^{\alpha-k}_{a+} y)(a+) = b_k, \quad k = 1, \ldots, n,$$

called also Riemann–Liouville FDE.

Definition 2 Let the FDE

$$(D^{\alpha}_{a+} y)(t) = f[t, y(t)], \quad \alpha > 0, \quad t > a,$$

with the initial conditions:

$$(D^k y)(0) = b_k, \quad k = 0, 1, \ldots, n-1,$$

called also Caputo FDE.

Lemma 1 *Let $y(t)$ be a function with continuous derivative in the interval $I_h(0) = [0, h]$ with values in $[y_0 - \eta, y_0 + \eta]$, then $y(t)$ satisfies the Caputo type*

$$D^{\alpha} y(t) = f(t, y(t)), \quad 0 < \alpha \leq 1, \qquad t > 0,$$

$$y(0) = y_0,$$

C. Milici et al., *Introduction to Fractional Differential Equations*, Nonlinear Systems and Complexity 25, https://doi.org/10.1007/978-3-030-00895-6_4

if and only if it satisfies the Voltera[1] *integral,*

$$y(t) = y_0 + \frac{1}{\Gamma(\alpha)} \int_0^t (t-u)^{\alpha-1} f(u, y(u))\, du.$$

Proof Let $L[y(t] = Y$ be the LT of $y(t)$. We have

$$s^\alpha Y - s^{\alpha-1} y_0 = L[f(t, y(t))],$$

$$Y = \frac{y_0}{s} + \frac{1}{s^\alpha} L[f(t, y(t))],$$

$$y(t) = y_0 + \frac{1}{\Gamma(\alpha)} \int_0^t (t-u)^{\alpha-1} f(u, y(u))\, du.$$

Definition 3 (Chebyshev[2] Norm) The Chebyshev norm on a set S is:

$$\|f\|_\infty = \sup\{|f(x)| : x \in S\},$$

where Supremum (sup) denotes the supremum.

Lemma 2 (The Weierstrass Test) *Suppose that* $\{f_n(t)\}$ *is a sequence of real functions defined on a set A, and there is a sequence of positive numbers* $\{R_n\}$ *satisfying:*

$$\forall n > 1, \quad \forall t \in A, \qquad |f_n(t)| \le R_n, \qquad \sum_{n=1}^{\infty} R_n < \infty.$$

Then the series $\sum_{n=1}^{\infty} f_n(t)$ *is convergent.*

Theorem 1 (Existence and Uniqueness for the Caputo Problem) *Let a Caputo FDE be*

$$D^\alpha y(t) = f(t, y(t)), \qquad 0 < \alpha \le 1, \qquad t > 0,$$

with the initial condition:

$$y(0) = y_0.$$

We consider the domain

$$D = [0, \eta] \times [y_0 - \eta, y_0 + \eta],$$

[1] V. Voltera (1860–1940).

[2] P.L. Chebyshev (1821–1894).

on which f satisfies:

- $f(t, y)$ *is continuous,*
- $|f(t, y)| < M$, *where* $M = \max_{(t,y)\in D} |f(t, y)|$, *and Maximum* (max) *denotes the maximum function*
- $f(t, y)$ *satisfy in D the Lipschitz*[3] *condition in y if there is a constant K such that:*

$$|f(t, y_2) - f(t, y_1)| \leq K|y_2 - y_1|.$$

Then it exists $\delta > 0$ and a function $y(t) \in C[0, \eta]$ unique for

$$\delta = \min\left\{\eta, \left(\frac{\eta\Gamma(\alpha+1)}{M}\right)^{1/\alpha}\right\},$$

where Minimum (min) *denotes the minimum function.*

Proof We consider the Voltera integral (see Lemma 1)

$$y(t) = y_0 + \frac{1}{\Gamma(\alpha)}\int_0^t (t-u)^{\alpha-1} f(u, y(u))\, du,$$

and successive approximations:

$$y_n(t) = y_0 + \frac{1}{\Gamma(\alpha)}\int_0^t (t-u)^{\alpha-1} f(u, y_{n-1}(u))\, du.$$

Using the method of successive approximations, on the basis of the Weierstrass test we prove the existence and the uniqueness of the solution of Caputo FDE.

For the sequence $\{y_n(t)\}$, we can prove that:

(i) the sequence $\{y_n(t)\}$ is well defined,
(ii) the sequence is uniformly continuous,
(iii) and its limit $y(t)$ is unique.

Proof

(i) We will use the induction method. In the case $n = 0$ it is obvious.
If $n = 1$, then we have:

$$|y_1(t) - y_0| = \left|\frac{1}{\Gamma(\alpha)}\int_0^t (t-u)^{\alpha-i} f(u, y_0)\, du\right| < \left|\frac{M}{\Gamma(\alpha)}\int_0^t (t-u)^{\alpha-1}\, du\right|$$
$$= \left|\frac{Mt^\alpha}{\Gamma(\alpha+1)}\right| \leq \left|\frac{M\delta^\alpha}{\Gamma(\alpha+1)}\right| < \eta.$$

---

[3]R.O.S. Lipschitz (1832–1903).

If we assume

$$|y_{n-1} - y_0| \leq \eta,$$

then it follows that:

$$|y_n - y_0| = \left|\frac{1}{\Gamma(\alpha)}\int_0^t (t-u)^{\alpha-1} f(u, y_{n-1})du\right| < \left|\frac{M}{\Gamma(\alpha)}\int_0^t (t-u)^{\alpha-1}\,du\right|$$
$$= \left|\frac{Mt^\alpha}{\Gamma(\alpha+1)}\right| \leq \left|\frac{M\delta^\alpha}{\Gamma(\alpha+1)}\right| < \eta.$$

(ii) We consider the series

$$y_0 + \sum_{k=0}^{\infty}(y_{k+1}(t) - y(t)),$$

equal with:

$$y_0 + \sum_{k=0}^{n-1}(y_{k+1}(t) - y_k(t)) = y_{n+1}(t).$$

We have:

$$|y_2 - y_1| = \left|\frac{1}{\Gamma(\alpha)}\int_0^t (t-u)^{\alpha-1}[f(u, y_1(u)) - f(u, y_0)]\,du\right|$$
$$\leq \left|\frac{K}{\Gamma(\alpha)}\int_0^t (t-u)^{\alpha-1}|y_1 - y_0|du\right| \leq \left|\frac{KM}{\Gamma(\alpha+1)}t^\alpha\right|$$
$$\leq \left|\frac{KM}{\Gamma(\alpha+1)}\delta^\alpha\right| = K\eta.$$

$$|y_3 - y_2| = \left|\frac{1}{\Gamma(\alpha)}\int_0^t (t-u)^{\alpha-1}[f(u, y_2(u)) - f(u, y_1)]du\right|$$
$$\leq \left|\frac{K}{\Gamma(\alpha)}\int_0^t (t-u)^{\alpha-1}|y_2 - y_1|du\right| \leq \left|\frac{K^2\eta}{\Gamma(\alpha+1)}t^\alpha\right|$$
$$\leq \left|\frac{K^2\eta}{\Gamma(\alpha+1)}\delta^\alpha\right| = \frac{K^2\eta^2}{M},$$

$$|y_{n+1} - y_n| \leq \frac{K^n\eta^n}{M^{n-1}}.$$

Using the Weierstrass test [9, 10] we obtain:

$$\sum_{n=0}^{\infty} |\ldots| = y_0 + \eta + \frac{1}{M} \sum_{n=1}^{\infty} \left(\frac{K\eta}{M}\right)^n.$$

The series are convergent for $\eta < \frac{M}{K}$.

Thus the sequence $\{y_n(t)\}$ is uniform convergent on the compact $[0, \eta]$. Hence, $y_n(t)$ is convergent to a function $y(t)$ for $t \in [0, \eta]$.

$\forall \eta > 0, \exists N$ positive number so for $n > N$ we have:

$$|y_n(t) - y(t)| < \eta.$$

This limit is unique.

(iii) Let $x(t)$ be another limit for $\{y_n(t)\}$, then:

$$\begin{aligned} |x(t) - y(t)| &= |x(t) - y_n(t) + y_n(t) - y(t)| \\ &\leq |y_n(t) - x(t)| + |y_n(t) - y(t)| \leq \frac{\eta}{2} + \frac{\eta}{2} = \eta. \end{aligned}$$

Remark (Another Solution) In order to prove the existence of the solution we can introduce the set

$$U = \{y \in C[0, \eta] : \|y - y_0\| \leq \eta\}$$

and an operator A:

$$Ay(t) = y_0 + \frac{1}{\Gamma(\alpha)} \int_0^t (t-u)^{\alpha-1} f(u, y(u))\, du,$$

where A has a fixed point, and U is a closed and convex subset of all continuous functions on $[0, \eta]$ equipped with Chebyshev norm [12].

Generally:

$$y(t) = \sum_{j=0}^{n} \frac{b_j}{\Gamma(\alpha - j + 1)} (t-a)^{\alpha-j} + \frac{1}{\Gamma(\alpha)} \int_0^t \frac{f(u, y(u))}{(t-u)^{1-\alpha}}\, du,$$

where $t > 0,\ n - 1 \leq \alpha < n$.

The technique used for proving the existence solution of the Voltera equation is often the successive approximation:

$$y_0(t) = \sum_{k=1}^{n} \frac{b_k}{\Gamma(\alpha - k + 1)} t^{\alpha-k},$$

$$y_i(t) = y_0(t) + \frac{1}{\Gamma(\alpha)} \int_0^t (t-\tau)^{\alpha-1} f(\tau, y_{i-1}(\tau))d\tau, \qquad i = 1, 2, \ldots$$

$$y(t) = \lim_{i\to\infty} y_i(t).$$

Example 1 Using method of successive approximation we solve the FDE:

$$D^\alpha y(t) = t^2 + y^2, \quad 0 < \alpha \le 1, \quad y(0) = 0, \quad (t, y) \in [-1, 1] \times [-1, 1].$$

Solution

- For $\alpha = 1$, we have method of successive approximation or the method of Picard.[4]

  We construct a sequence $\{y_n(t)\}$ by the recurrence

$$y_n(t) = y_0 + \int_0^t f[u, y_{n-1}(u)]du, \quad n = 1, 2, \ldots$$

  The $\{y_n(t)\}$ is convergent to an exact solution of the equation

$$y'(t) = f[t, y(t)] = t^2 + y^2, \quad y(0) = 0,$$

  in some interval $0 - h < t < 0 + h$ in the rectangle $|t - t_0| \le a = 1$, $|y - y_0| \le b = 1$,

$$h = \min\left(a, \frac{b}{M}\right), \quad M = \max_{(t,y)\in D} |f(t, y)|,$$

  $y_n(t)$ is given by the inequality

$$|y(t) - y_n(t)| \le \frac{MN^{n-1}}{n!} h^n, \quad N = \max_{(t,y)\in D} \left|\frac{\partial f}{\partial y}\right|.$$

  For

$$M = 2, \quad a = 1, \quad h = \frac{1}{2},$$

  it results:

$$y_0(t) = 0,$$
$$y_1(t) = \int_0^t (u^2 + y_0^2)du = \frac{t^3}{3},$$

[4] E. Picard (1856–1941).

$$y_2(t) = \int_0^t (u^2 + y_1^2)du = \frac{t^3}{3} + \frac{t^7}{63},$$
$$y_3(t) = \int_0^t (u^2 + y_1^2)du = \frac{t^3}{3} + \frac{t^7}{63} + \frac{2t^11}{2079} + \frac{t^15}{59535},$$

$$|y_3(t) - y(t)| \leq \frac{2}{3!}\left(\frac{1}{2}\right)^3 2^2 = \frac{1}{6}, \quad N = \max|2y| = 2.$$

- For $0 < \alpha \leq 1$, we obtain:

$$y_0 = 0,$$

$$y_n(t) = \frac{1}{\Gamma(\alpha)} \int_0^t (t-u)^{\alpha-1}[u^2 + y_{n-1}^2(u)]du.$$

We can calculate $y_1(t)$:

$$y_1(t) = \frac{1}{\Gamma(\alpha)} \int_0^t (t-u)^{\alpha-1} u^2 du.$$

The LT of this convolution is

$$Y_1 = \frac{1}{\Gamma(\alpha)} L[u^{\alpha-1}]L[u^2] = \frac{1}{\Gamma(\alpha)} \frac{\Gamma(\alpha)}{s^\alpha} \frac{\Gamma(3)}{s^3} = \frac{\Gamma(3)}{s^{\alpha+3}},$$

from which, by inversion, we obtain:

$$y_1(t) = \frac{2t^{\alpha+2}}{\Gamma(\alpha+3)}.$$

Similarly, we obtain also:

$$y_2(t) = \frac{1}{\Gamma(\alpha)} \int_0^t (t-u)^{\alpha-1} \left[u^2 + 4\frac{u^{2\alpha+4}}{\Gamma^2(\alpha+3)}\right] du,$$

with:

$$Y_2(s) = \frac{2}{s^{\alpha+3}} + \frac{4}{\Gamma^2(\alpha+3)} \frac{\Gamma(2\alpha+5)}{s^{3\alpha+5}},$$

and finally:

$$y_2(t) = \frac{2t^{\alpha+2}}{\Gamma(\alpha+3)} + \frac{4\Gamma(2\alpha+5)}{\Gamma^2(\alpha+3)} \frac{t^{3\alpha+4}}{\Gamma(3\alpha+5)},$$

For $\alpha = 1$, we obtain $y_2(t) = \frac{t^3}{3} + \frac{t^7}{63}$.

Example 2 Using method of successive approximation we solve the FDE:

$$D^\alpha y(t) = 1 + ty(t) + y^2(t), \qquad 0 < \alpha \leq 1, \qquad y(0) = 0.$$

Solution As in the previous example, we have:

$$y_0(t) = 0,$$
$$y_n(t) = \frac{1}{\Gamma(\alpha)} \int_0^t (t-u)^{\alpha-1} \left[1 + uy_{n-1}(u) + y_{n-1}^2(u)\right] du,$$
$$y_1(t) = \frac{1}{\Gamma(\alpha)} \int_0^t (t-u)^{\alpha-1}\, du,$$
$$Y_1 = \frac{1}{\Gamma(\alpha)} L[u^{\alpha-1}]L[1] = \frac{1}{\Gamma(\alpha)} \frac{\Gamma(\alpha)}{s^\alpha} \frac{1}{s} = \frac{\Gamma(1)}{s^{\alpha+1}},$$
$$y_1(t) = \frac{t^\alpha}{\Gamma(\alpha+1)}$$

$$y_2(t) = \frac{1}{\Gamma(\alpha)} \int_0^t (t-u)^{\alpha-1} \left[1 + u \cdot y_1(u) + y_1^2(u)\right] du$$

$$Y_2 = \frac{1}{s^{\alpha+1}} \Gamma(\alpha+1) + \frac{\Gamma(2\alpha+1)}{\Gamma^2(\alpha+1)} \frac{1}{s^{3\alpha+1}} + \frac{\alpha+1}{s^{2\alpha+2}},$$

$$y_2 = \frac{t^\alpha}{\Gamma(\alpha+1)} + \frac{\Gamma(2\alpha+1)}{\Gamma^2(\alpha+1)} \frac{t^{3\alpha}}{\Gamma(3\alpha+1)} + (\alpha+1)\frac{t^{2\alpha+1}}{\Gamma(2\alpha+1)}.$$

4.2 Linear Fractional Differential Equations

A **linear** FDE is an equation of form

$$(D^{\alpha_n} + a_{n-1}D^{\alpha_{n-1}} + \ldots a_1 D^{\alpha_1} + a_0)y(t) = f(t), \qquad \alpha \in \mathbb{R},$$

with the conditions:

$$y^{(k)}(0) = b_k, \qquad k = 0, 1, 2, \ldots, n-1.$$

An equation which is not linear is called nonlinear.

Theorem 2 (Existence and Uniqueness) *If $f(t)$ is bounded on $(0, T)$ and $a_k = a_k(t)$, $k \in \{0, 1, \ldots, n-1\}$ are continuous functions on $[0, T]$, the equation has a unique solution.*

Proof The proof used here will be based on the proof of the existence and uniqueness of the solution of Voltera integral equation.

Theorem 3 *The linear FDE:*

$$D^{\alpha} y(t) = f(t), \qquad \text{where: } n-1 < \alpha < n,$$

and

$$y^{(k)}(0) = 0, \qquad k = 0, 1, 2, \ldots, n-1,$$

has the solution:

$$y(t) = \frac{1}{\Gamma(\alpha)} \int_0^t (t-u)^{\alpha-1} f(u)\, du.$$

Proof We apply the LT:

$$L[y(t)] = Y = Y(s),$$

$$L\left[D^{\alpha} y(t)\right] = L[f(t)] = F(s),$$

$$s^{\alpha} Y = F(s) \quad \Rightarrow \quad Y = \frac{F(s)}{s^{\alpha}}$$

and using the convolution theorem it results the assumption of this theorem.

Theorem 4 *The linear FDE:*

$$D^{\alpha} y(t) = \lambda y(t), \qquad \text{where: } n-1 < \alpha < n,$$

with the initial condition

$$y^{(k)}(0) = b_k, \qquad b_k \in R, \qquad k = 0, 1, 2, \ldots, n-1,$$

has the solution:

$$y(t) = \sum_{k=0}^{n-1} b_k t^k E_{\alpha,k+1}(\lambda t^{\alpha}).$$

Proof We apply the LT method:

$$L[y(t)] = Y = Y(s),$$

$$L\left[D^{\alpha} y(t)\right] = \lambda L[y(t)],$$

$$s^{\alpha}Y - \sum_{k=0}^{n-1} s^{\alpha-k-1} y^{(k)}(0) - \lambda Y = 0,$$

$$Y = \sum_{k=0}^{n-1} \frac{s^{\alpha-k-1}}{s^{\alpha}-\lambda} b_k = \sum_{k=0}^{n-1} L\left[b_k t^k E_{\alpha,k+1}(\lambda t^{\alpha})\right] = L\left[\sum_{k=0}^{n-1} b_k t^k E_{\alpha,k+1}(\lambda t^{\alpha})\right].$$

Because $L[y(t)] = Y$ it results the statements of the theorem.

Theorem 5 *The linear FDE:*

$$D^{\alpha} y(t) = f(t), \qquad \textit{where: } 0 < \alpha < 1,$$

with the initial condition

$$y(0) = A,$$

and where:

$$f(t) = \sum_{n=0}^{\infty} \frac{f^{(n)}(0)}{n!} t^n,$$

has the solution:

$$y(t) = A + t^{\alpha} \sum_{n=0}^{\infty} \frac{f^{(n)}(0)}{\Gamma(n+\alpha+1)} t^n.$$

Proof We apply the LT:

$$L[y(t)] = Y = Y(s),$$

$$L\left[D^{\alpha} y(t)\right] = s^{\alpha} Y - A s^{\alpha-1},$$

$$s^{\alpha} Y - A s^{\alpha-1} = L[f(t)] = \sum_{n=0}^{\infty} \frac{f^{(n)}(0)}{n!} L[t^n],$$

$$y(t) = \sum_{n=0}^{\infty} y_n, \quad Y = L[y(t)] = \sum_{n=0}^{\infty} Y_n,$$

$$\sum_{n=0}^{\infty} Y_n = \frac{A}{s} + \frac{1}{s^{\alpha}} \sum_{n=0}^{\infty} \frac{f^{(n)}(0)}{n!} L[t^n],$$

$$Y_0 = \frac{A}{s} \qquad \Rightarrow y_0 = A,$$

$$Y_n = \frac{f^{(n)}(0)}{\Gamma(n+1)}\frac{\Gamma(n+1)}{s^{n+\alpha+1}} \quad \Rightarrow y_n = \frac{f^{(n)}(0)}{\Gamma(n+\alpha+1)},$$

$$y(t) = A + \sum_{n=0}^{\infty} \frac{f^{(n)}(0)}{\Gamma(n+\alpha+1)} t^{n+\alpha}.$$

Theorem 6 *The linear FDE:*

$$aD^{\alpha}y(t) + by(t) = f(t), \qquad 0 < \alpha < 1,$$

$$y(0) = A,$$

where:

$$f(t) = \sum_{n=0}^{\infty} \frac{f^{(n)}(0)}{n!} t^n,$$

has the solution:

$$y(t) = AE_{\alpha,1}\left(-\frac{b}{a}t^{\alpha}\right) + \sum_{n=0}^{\infty} \frac{f^{(n)}(0)}{a} t^n \left[1 - E_{\alpha,n+1}(-\frac{b}{a}t^{\alpha})\right].$$

Proof Applying the LT, it results:

$$L[y(t)] = Y = Y(s),$$

$$L\left[D^{\alpha}y(t)\right] = s^{\alpha}Y - As^{\alpha-1},$$

$$s^{\alpha}Y - As^{\alpha-1} + bY = L[f(t)] = \sum_{n=0}^{\infty} \frac{f^{(n)}(0)}{n!} L[t^n],$$

$$y(t) = \sum_{n=0}^{\infty} y_n, \qquad Y = L[y(t)] = \sum_{n=0}^{\infty} Y_n,$$

$$\sum_{n=0}^{\infty} Y_n = \frac{As^{\alpha-1}}{as^{\alpha}+b} + \frac{1}{as^{\alpha}+b}\sum_{n=0}^{\infty} \frac{f^{(n)}(0)}{n!} L[t^n],$$

$$Y_0 = \frac{As^{\alpha-1}}{as^\alpha + b} \quad \Rightarrow \quad Y_0 = \frac{A}{as}\frac{1}{1+\dfrac{b}{as^\alpha}},$$

$$Y_n = \frac{1}{as^\alpha + b}\frac{f^{(n)}(0)}{n!}L[t^n] \;\Rightarrow\; Y_n = \frac{f^{(n)}(0)}{as^{n+1+\alpha}}\frac{1}{1+\dfrac{b}{as^\alpha}},$$

$$Y_0 = A\left[\frac{1}{s} - \frac{b}{a}\frac{1}{s^{\alpha+1}} + \frac{b^2}{a^2}\frac{1}{s^{2\alpha+1}} + \ldots\right],$$

$$y_0 = A\left[1 - \frac{b}{a}\frac{t^\alpha}{\Gamma(\alpha+1)} + \frac{b^2}{a^2}\frac{t^{2\alpha}}{\Gamma(2\alpha+1)} + \ldots\right],$$

$$y_0 = AE_{\alpha,1}\left(-\frac{b}{a}t^\alpha\right),$$

$$Y_n = \frac{f^{(n)}(0)}{a}\left[\frac{1}{s^{n+1+\alpha}} - \frac{b}{a}\frac{1}{s^{n+1+2\alpha}} + \frac{b^2}{a^2}\frac{1}{s^{n+1+3\alpha}} + \ldots\right],$$

$$y_n = \frac{f^{(n)}(0)}{a}\left[\frac{t^{n+\alpha}}{\Gamma(n+1+\alpha)} - \frac{b}{a}\frac{t^{n+2\alpha}}{\Gamma(n+1+2\alpha)} + \frac{b^2}{a^2}\frac{t^{n+3\alpha}}{\Gamma(n+1+3\alpha)} + \ldots\right],$$

$$y_n = \frac{f^{(n)}(0)}{a}t^n\left[1 - E_{\alpha,n+1}\left(-\frac{b}{a}t^\alpha\right)\right].$$

Finally, we obtain the solution:

$$y(t) = y_0 + \sum_{n=0}^{\infty} y_n,$$

$$y(t) = AE_{\alpha,1}\left(-\frac{b}{a}t^\alpha\right) + \sum_{n=0}^{\infty}\frac{f^{(n)}(0)}{a}t^n\left[1 - E_{\alpha,n+1}\left(-\frac{b}{a}t^\alpha\right)\right].$$

Example 1 We will establish here the solution of the FDE:

$$D^\alpha y(t) + y(t) = 1,$$

with the initial condition:

$$y(0) = 0.$$

Solution We apply the LT:

$$L\left[D^{\alpha}y(t)\right] + L[y(t)] = L[1],$$

$$s^{\alpha}Y - s^{\alpha-1}y(0) + Y = \frac{1}{s},$$

$$Y = \frac{1}{s(s^{\alpha}+1)},$$

$$Y = \frac{1}{s^{\alpha+1}}\frac{1}{1+\frac{1}{s^{\alpha}}},$$

and using the identity

$$\frac{1}{1+u} = 1 - u + u^2 - u^3 + \ldots \qquad |u| < 1,$$

we obtain:

$$Y = \frac{1}{s^{\alpha+1}} - \frac{1}{s^{2\alpha+1}} + \frac{1}{s^{3\alpha+1}} + \ldots$$

Finally, it results:

$$y(t) = \frac{t^{\alpha}}{\Gamma(\alpha+1)} - \frac{t^{2\alpha}}{\Gamma(2\alpha+1)} + \frac{t^{3\alpha}}{\Gamma(3\alpha+1)} + \ldots$$

For $\alpha = 1$, we can use the Maple or Mathematica commands in order to establish the solution:

MAPLE

```
ec: = diff(y(t),t) + y(t) = 1;
dsolve({ec,y(0) = 0},y(t),type = series);
```

MATHEMATICA

```
Clear["`*"]c
ec := y'[t] + y[t] == 1;
sol = DSolve[{ec, y[0] == 0}, y, t]
Series[y[t] /. sol, {t, 0, 10}]
```

It results the solution:

$$y(t) = t - \frac{t^2}{3} + \frac{t^3}{6} - \frac{t^4}{24} + \ldots$$

Example 2 We consider the FDE

$$D^\alpha y(t) = y(t),$$

with the initial conditions:

$$y(0) = 0, \qquad y'(0) = 1.$$

We will establish the solution of this equation for the cases:

1. $1 < \alpha \leq 2$,
2. $2 < \alpha \leq 3$.

Solution

1. For the case $1 < \alpha \leq 2$, using the LT, we have:

$$L[D^\alpha y(t]) = L[y(t)],$$

$$s^\alpha Y - s^{\alpha-1} y(0) - s^{\alpha-2} y'(0) = Y,$$

$$Y = \frac{s^{\alpha-2}}{s^\alpha - 1},$$

$$Y = \frac{1}{s^2} \frac{1}{1 - \frac{1}{s^\alpha}},$$

and using the identity

$$\frac{1}{1-u} = 1 + u + u^2 + u^3 + \qquad |u| < 1,$$

we have

$$Y = \frac{1}{s^2} + \frac{1}{s^{\alpha+2}} + \frac{1}{s^{2\alpha+2}} + \ldots,$$

and the solution:

$$y(t) = \frac{t}{\Gamma(2)} + \frac{t^{\alpha+1}}{\Gamma(\alpha+2)} + \frac{t^{2\alpha+1}}{\Gamma(2\alpha+2)} + \ldots$$

We can note that:

$$\lim_{\alpha \to 2} y(t) = \frac{t}{1!} + \frac{t^3}{3!} + \frac{t^5}{5!} \ldots = \sinh(t).$$

2. For the case $2 < \alpha \leq 3$, using same procedure, we have:

$$s^{\alpha}Y(s) - s^{\alpha-1}y(0) - s^{\alpha-2}y'(0) - s^{\alpha-3}y''(0) = Y(s).$$

For $y''(0) = b$, we obtain:

$$Y = \frac{s^{\alpha-2}}{s^{\alpha}-1} + b\frac{s^{\alpha-3}}{s^{\alpha}-1},$$

and using the residues theorem we have:

$$\lim_{\alpha \searrow 2} Y = \frac{s+b}{s(s^2-1)},$$

$$r_1 = \operatorname*{Res}_{0}(Ye^{st}) = \lim_{s\to 0} sYe^{st} = -b,$$

$$r_2 = \operatorname*{Res}_{-1}(Ye^{st}) = \lim_{s\to -1}(s+1)Ye^{st} = \frac{(b-1)e^{-t}}{2},$$

$$r_3 = \operatorname*{Res}_{1}(Ye^{st}) = \lim_{s\to 1}(s-1)Ye^{st} = \frac{(b+1)e^{t}}{2}.$$

The solution will be:

$$y(t) = r_1 + r_2 + r_3 = \sinh(t) + b\cosh(t) - b.$$

Observation

This equation can be solved also in terms of perturbation method. In this case we take

$$y(t) = t + c_1 t^{\alpha+1} + c_2 t^{2\alpha+1} + c_3 t^{3\alpha+1} + \ldots,$$

and using the formula:

$$D_*^{\alpha} t^{\beta} = \begin{cases} \dfrac{\Gamma(\beta+1)}{\Gamma(\beta+1-\alpha)} t^{\beta-\alpha} & \beta > n-1 \\ 0 & \beta \leq n-1. \end{cases}$$

$$\begin{aligned} D_*^{\alpha} y(t) &= c_1 D_*^{\alpha} t^{\alpha+1} + c_2 D_*^{\alpha} t^{2\alpha+1} + c_3 D_*^{\alpha} t^{\alpha+1} + \ldots \\ &= t + c_1 t^{\alpha+1} + c_2 t^{2\alpha+1} + c_3 t^{3\alpha+1} + \ldots \end{aligned}$$

we have

$$c_1\frac{\Gamma(\alpha+2)}{\Gamma(2)}t+c_2\frac{\Gamma(2\alpha+2)}{\Gamma(\alpha+2)}t^{\alpha+1}+c_3\frac{\Gamma(3\alpha+2)}{\Gamma(2\alpha+2)}t^{2\alpha+1}+\dots$$
$$=t+c_1t^{\alpha+1}+c_2t^{2\alpha+1}+c_3t^{3\alpha+1}+\dots$$

and after the identification we obtain:

$$c_1\Gamma(\alpha+2)=1 \qquad \Rightarrow \qquad c_1=\frac{1}{\Gamma(\alpha+2)},$$

$$c_2\frac{\Gamma(2\alpha+2)}{\Gamma(\alpha+2)}=c_1 \qquad \Rightarrow \qquad c_2=\frac{1}{\Gamma(2\alpha+2)},$$

$$c_3\frac{\Gamma(3\alpha+2)}{\Gamma(2\alpha+2)}=c_2 \qquad \Rightarrow \qquad c_3=\frac{1}{\Gamma(3\alpha+2)},$$

$$\dots$$

For $\alpha=2$, we have:

$$y''(t)=y(t).$$

We apply the LT:

$$L[y''(t)]=L[y(t)], \qquad L[y(t)]=Y=Y(s),$$

$$s^2Y-sy'(0)-y(0)=Y,$$

$$Y=\frac{1}{s^2-1},$$

$$r_1=\operatorname*{Res}_{-1}Ye^{st}=\lim_{s\to-1}Ye^{st}=\lim_{s\to-1}(s+1)\frac{e^{st}}{(s-1)(s+1)}=-\frac{e^{-t}}{2},$$

$$r_2=\operatorname*{Res}_{1}Ye^{st}=\lim_{s\to1}Ye^{st}=\lim_{s\to1}(s-1)\frac{e^{st}}{(s-1)(s+1)}=\frac{e^{t}}{2},$$

$$f(t)=r_1+r_2=\frac{e^t-e^{-t}}{2}=\sinh(t).$$

We can use here the **MAPLE** commands:

```
with(inttrans);
ec:=diff(y(t),t$2) = y(t);
dsolve(ec,D(y)(0) = 1,y(0) = 0,y(t), method = laplace);
```

Example 3 Find the solution of the FDE:

$$D^2 y(t) - D^{3/2} y(t) - y(t) + t + 1 = 0,$$

with the initial conditions:

$$y(0) = y'(0) = 1.$$

Solution We apply the LT:

$$L\left[D^2 y(t)\right] - L\left[D^{\frac{3}{2}} y(t)\right] - L[y(t)] + L[t] + L[1] = 0,$$

$$L\left[D^2 y(t)\right] = s^2 Y - s y(0) - y'(0) = s^2 Y - s - 1,$$

$$L\left[D^{\frac{3}{2}} y(t)\right] = s^{\frac{3}{2}} Y - s^{\frac{1}{2}} y(0) - s^{-\frac{1}{2}} y'(0) = s^{\frac{1}{2}} Y - s^{\frac{1}{2}} - s^{-\frac{1}{2}} =$$

$$= \frac{s^2 - s - 1}{\sqrt{s}},$$

$$L[y(t)] = Y,$$

$$L[t] = \frac{1}{s^2},$$

$$L[1] = \frac{1}{s},$$

$$s^2 Y - s - 1 - \frac{s^2 Y - s - 1}{\sqrt{s}} - \left[s^2 Y - s - 1\right] = 0,$$

$$s^2 Y - s - 1 = 0, \qquad \Rightarrow \qquad Y = \frac{1}{s} + \frac{1}{s^2},$$

$$y(t) = L^{-1}[Y], \qquad \Rightarrow \qquad y(t) = 1 + t.$$

Example 4 We consider the problem [11], with the initial condition:

$$\frac{d^{1/2} y}{dx^{1/2}} = \frac{y}{x}, \qquad y(0) = 0.$$

which can be rewritten as:

$$t D^{1/2} y(t) - y(t) = 0, \qquad y(0) = 0.$$

We will establish the solution of this equation.

Solution We apply the LT method:

$$L\left[tD^{1/2}y(t)\right] - L[y(t)] = 0,$$

$$L\left[tD^{1/2}y(t)\right] = -\frac{d}{ds}\left[s^{1/2}Y - s^{-1/2}y(0)\right] - Y = 0,$$

$$\frac{dY}{ds} + \left(\frac{1}{2s} + \frac{1}{\sqrt{s}}\right)Y = 0.$$

We obtain:

$$Y(s) = C\frac{e^{-2\sqrt{s}}}{\sqrt{s}} \qquad \Rightarrow \qquad y(t) = \frac{Ce^{-1/t}}{\sqrt{\pi t}}.$$

The same solution can be found using the Maple program:

MAPLE

```
with(inttrans):
ec:= diff(Y(s),s) + (1/(2*s)+1/sqrt(s)) - Y(s) = 0;
F(s):= dsolve(ec,Y(s));
f(t):= invlaplace(F(s),s,t);
```

Other applications can be found in [5, 6].

4.3 Nonlinear Equations

4.3.1 The Adomian Decomposition Method

The Adomian[5] method [1–3], applied to the ordinary and partial differential equations of integer order was extended also to the case of FDE (for further details and examples see [7, 8]).

Adomian Polynomials

We will denote these polynomials by $A_0, A_1, \ldots A_n, \ldots$.

We consider a nonlinear analytic function $G(y(t), t)$ and that $y(t_0) = y_0$, in the D domain. The Adomian method consists in the decomposition the unknown function $y(t)$ in a series of form $y(t) = y_0 + y_1 + y_2 + \ldots + y_n + \ldots$, where y_n can be expressed in terms of Adomian polynomials A_n.

[5]G. Adomian (1922–1996).

The Adomian polynomials are defined [1, 2, 8]:

$$A_n = \frac{1}{n!}\frac{d^n}{d\lambda^n}\left[G(t, \sum_{j=0}^{n} y_j\lambda^j)\right]_{\lambda=0}.$$

These polynomials can be established algorithmically, using the symbolic programming packages, with the aid of **do ... end do** repetition statements:

Step 1: To calculate first the Adomian polynomial A_0, i.e., $A_0 = G(y_0, t)$.
Step 2: Iterative calculation of A_k using the **for ... do .. end do** loop:

> for $k = 0$ to n - 1 do
> $A_k = A_k(y_0 + \lambda * y_1, \ldots, y_k + (k+1) * \lambda * y_{k+1})$
> end do;

It must be underlined that in the A_k polynomial y_i is replaced with

$$y_i \rightarrow y_i + (i+1) * y_{i+1} * \lambda, \qquad \text{for: } i = 0, 1, \ldots, k.$$

Step 3:

$$\frac{d}{d\lambda}A_k\bigg|_{\lambda=0} = (k+1) * A_k.$$

Step 4: We obtain, finally $A_0, A_1, \ldots, A_n$ and $y_0, y_1, y_2, \ldots$ in terms of A_n.

A given $f(u)$ can be expressed as a series of A_n,

$$f(u) = \sum_{n=0}^{\infty} A_n,$$

and $u = \sum_{n=0}^{\infty} u_n$.

The series $\sum_{n=0}^{\infty} A_n$ can be rearranged as a generalized Taylor series:

$$\begin{aligned} f(u) = \sum_{n=0}^{\infty} A_n &= f(u_0) + (u_1 + u_2 + \ldots) f^{(1)}(u_0) \\ &+ \left[\frac{u_1^2}{2!} + u_1u_2 + \ldots\right] f^{(2)}(u_0) + \ldots \\ &= \sum_{n=0}^{\infty} \left[(u-u_0)^n/n!\right] f^{(n)}(u_0) = \sum_{n=0}^{\infty} \left[(u_1+u_2+\ldots)^n/n!\right] f^{(n)}(u_0), \end{aligned}$$

so that:

$$A_0 = f(u_0),$$

$$A_1 = u_1 f^{(1)}(u_0),$$

$$A_2 = u_2 f^{(1)}(u_0) + (1/2!)u_1^2 f^{(2)}(u_0),$$

$$A_3 = u_3 f^{(1)}(u_0) + u_1 u_2 f^{(2)}(u_0) + (1/3!)u_1^3 f^{(3)}(u_0),$$

$$\ldots$$

Example 1 We will establish the Adomian polynomials for $G(y) = y^2$.
The Adomian polynomials will be:

$A_0 = y_0^2$, $A_1 = 2y_0y_1$, $A_2 = y_1^2 + 2y_0y_2$, $A_3 = 2y_1y_2 + 2y_0y_3$, etc.
For calculation you can use also the following Maple, or Mathematica sequences:

MAPLE

```
restart;
with(LinearAlgebra):
unassign('y,lambda'):
f:=y->y^2:
S:=lambda->sum(y[i]*lambda^i,i=0..4):
g:=lambda->(S(lambda))^2:
c:=Vector(4,n->diff(1/n!*g(lambda),lambda$n):
A:=<subs(lambda=0,g(lambda)),subs(lambda=0,c)>;
```

MATHEMATICA

```
Clear["`*"]
f[y_] := y^2;
S[\[Lambda]_] := Sum[y[i]*\[Lambda]^i, {i, 0, 5}];
g[\[Lambda]_] := f[S[\[Lambda]]];
ad = Table[
  1/n!*D[g[\[Lambda]], {\[Lambda],n}]/. \[Lambda]-> 0,
  {n, 0, 5}] // Simplify;
TableForm[ad, TableAlignments -> Left]
```

Example 2 Let us calculate the Adomian polynomials for $G = y^3$.

Solution

$A_0 = y_0^3$, $A_1 = 3y_1y_0^2$, $A_2 = 3y_0^2y_2 + 3y_1^2y_0$,
$A_3 = y_1^3 + 6y_0y_1y_2 + 3y_0^2y_1$, etc.

The basic principles of this algorithm remain unchanged for other definitions of the function G.

Example 3 Let us calculate the Adomian polynomials for $G = f(u)$.
We obtain successively:

$$A_0 = f(u_0),\ A_1 = u_1(d/du_0)f(u_0),$$

$$A_2 = u_2(d/du_0)f(u_0) + (u_1^2)/2!)(d^2/du_0^2)f(u_0),$$

$$A_3 = u_3(d/du_0)f(u_0) + (u_1u_2)(d^2/du_0^2)f(u_0) + (u_1^3/3!)(d^3/du_0^3)f(u_0), \text{etc.}$$

Example 4 Find the Adomian polynomials for $G = \sin\theta$. It results:

$$A_0 = \sin\theta_0,\ A_1 = \theta_1\cos\theta_0,\ A_2 = -(\theta_1^2/2)\sin\theta_0 + \theta_2\cos\theta_0, \text{ etc.}$$

4.3.2 Decomposition of Nonlinear Equations

We consider the nonlinear FDE of type:

$$D^\alpha y(t) + Ry(t) + Ny(t) = f(t), \quad y^{(k)}(0) = c_k, k = 0, 1, \ldots, n-1, \quad \alpha > 0,$$

where N is a nonlinear operator, and Ry is a residual part of the equation.
We apply the LT to the equation. It follows:

$$L[D^\alpha y(t)] = s^\alpha Y - s^{\alpha-1}y(0) - s^{\alpha-2}y'(0) - \ldots - y^{(n-1)}(0) = s^\alpha Y - c,$$

$$c = s^{\alpha-1}y(0) + s^{\alpha-2}y'(0) - \ldots + y^{(n-1)}(0),$$

where c is a constant.
We use the following decomposition of $y(t)$

$$y(t) = \sum_{n=0}^{\infty} y_n(t),$$

with

$$Ny(t) = \sum_{n=0}^{\infty} A_n,$$

where A_n are Adomian polynomials:

$$A_n = \frac{1}{n!}\frac{d^n}{d\lambda^n}\left[N\left(\sum_{i=0}^{\infty}\lambda^i y_i\right)\right],$$

$$L\left[\sum_{n=0}^{\infty} y_n\right] = \frac{c}{s^\alpha}Y - \frac{1}{s^\alpha}L\left[R\sum_{n=0}^{\infty} y_n\right] - \frac{1}{s^\alpha}L\left[\sum_{n=0}^{\infty} A_n\right].$$

We have after calculations:

$$Y_0 = L[y_0] = \frac{c}{s^\alpha} + \frac{1}{s^\alpha}L\left[f(t)\right],$$

$$Y_1 = L[y_1] = -\frac{1}{s^\alpha}L[Ry_0] - \frac{1}{s^\alpha}L[A_0],$$

$$Y_2 = L[y_2] = -\frac{1}{s^\alpha}L[Ry_1] - \frac{1}{s^\alpha}L[A_1],$$

$$\ldots$$

$$Y_n = L[y_n] = -\frac{1}{s^\alpha}L[Ry_{n-1}] - \frac{1}{s^\alpha}L[A_{n-1}].$$

Example 1 Solve the nonlinear FDE using the Adomian decomposition method:

$$D^\alpha y(t) = t + y^2, \qquad 1 < \alpha \le 2,$$

$$y(0) = 0, \qquad y'(0) = 1.$$

Solution In order to solve the equation we apply the LT:

$$L[D^\alpha y(t)] = L\left[t + y^2\right],$$

$$L[y(t)] = Y,$$

$$L[D^\alpha y(t)] = s^\alpha Y - s^{\alpha-1}y(0) - s^{\alpha-2}y'(0) = s^\alpha Y - s^{\alpha-2},$$

$$Y = \frac{1}{s^2} + \frac{1}{s^\alpha}L\left[t + y^2\right],$$

so that, for the decomposition

$$y(t) = \sum_{n=0}^{\infty} y_n(t),$$

we obtain:

$$Y=\sum_{n=0}^{\infty}Y_n, \qquad t+y^2=\sum_{n=0}^{\infty}A_n,$$

where A_n are Adomian polynomials. We obtain:

$$A_0=t+y_0^2,\ \ A_1=2y_0y_1,\ \ A_2=y_1^2+2y_0y_2,\ \ A_3=2y_1y_2+2y_0y_2,\dots$$

$$\sum_{n=0}^{\infty}Y_n=\frac{1}{s^2}+\frac{1}{s^{\alpha}}\sum_{n=0}^{\infty}A_n,$$

$$Y_0=\frac{1}{s^2}\ \Rightarrow\ y_0=t,$$

$$A_0=t+t^2,$$

$$Y_1=\frac{1}{s^{\alpha}}L[A_0] \qquad \Rightarrow \qquad Y_1=\frac{1}{s^{\alpha}}L\left[t+t^2\right],$$

$$Y_1=\frac{1}{s^{\alpha+2}}+2\frac{1}{s^{\alpha+3}} \qquad \Rightarrow \qquad y_1=\frac{t^{\alpha+1}}{\Gamma(\alpha+2)}+2\frac{t^{\alpha+2}}{\Gamma(\alpha+3)},$$

$$A_1=2y_0y_1,$$

$$A_1=2\frac{t^{\alpha+2}}{\Gamma(\alpha+2)}+4\frac{t^{\alpha+3}}{\Gamma(\alpha+3)},$$

$$Y_2=\frac{1}{s^{\alpha}}L[A_1]\ \Rightarrow\ Y_2=\frac{2(\alpha+2)}{s^{2\alpha+3}}+\frac{4(\alpha+3)}{s^{2\alpha+4}},$$

$$y_2=\frac{2(\alpha+2)t^{2\alpha+2}}{\Gamma(2\alpha+3)}+\frac{4(\alpha+3)t^{2\alpha+3}}{\Gamma(2\alpha+4)},$$

$$y(t)=y_0(t)+y_1(t)+y_2(t)+\dots,$$

and finally, it yields:

$$y(t)=t+\frac{t^{\alpha+1}}{\Gamma(\alpha+2)}+2\frac{t^{\alpha+2}}{\Gamma(\alpha+3)}+\frac{2(\alpha+2)t^{2\alpha+2}}{\Gamma(2\alpha+3)}+\frac{4(\alpha+3)t^{2\alpha+3}}{\Gamma(2\alpha+4)}+\dots$$

For $\alpha=2$, we obtain:

$$y''(t)=t^2+y^2(t), \qquad y(0)=0, \qquad y'(0)=1.$$

The result

$$y(t) = t + \frac{t^3}{6} + \frac{t^4}{12} + \dots,$$

can be obtained also on computer, using the sequences:

MAPLE

```
ec: = diff(y(t), t$2) = t + (y(t))^2;
dsolve({ec,y(0) = 0,D(y)(0) = 1},y(t),type = series);
```

MATHEMATICA

```
ec:=y"[t] == t + y[t]*y[t];
sol=DSolve[{ec,y[0]==0,y'[0]==1},y,t];
Series[y[t]/.sol,{t,0,10}]
```

MATHEMATICA

```
Clear["`*"]
Manipulate[
 f[t_] := t + t^3/6 + t^4/12 + t^5/120;
 y[t_] :=
  t + t^(a + 1)/Gamma[a + 2] + 2*t^(a + 2)/Gamma
   [a + 3] + 2*(a + 2)*t^(2*a + 2)/Gamma[2*a + 3] +
   4*(a + 3)*t^(2*a + 3)/Gamma[2*a + 4];
Plot[{f[t], y[t]}, {t, 0, 1}, ImageSize -> 300,
  Frame -> True], {{a, 1/2}, 0, 1}]
```

The functions $f(t)$ and $y(t)$ calculated here are plotted in Fig. 4.1.

Example 2 Let us solve the FDE:

$$D^{\alpha} y(t) = 1 + y^2(t), \qquad 0 < \alpha \leq 1,$$

where:

$$y(0) = 0,$$

using the Adomian decomposition method.

Solution To solve this problem we apply the LT:

$$L[D^{\alpha} y(t)] = L[1] + L\left[y^2\right],$$

$$L[y(t)] = Y,$$

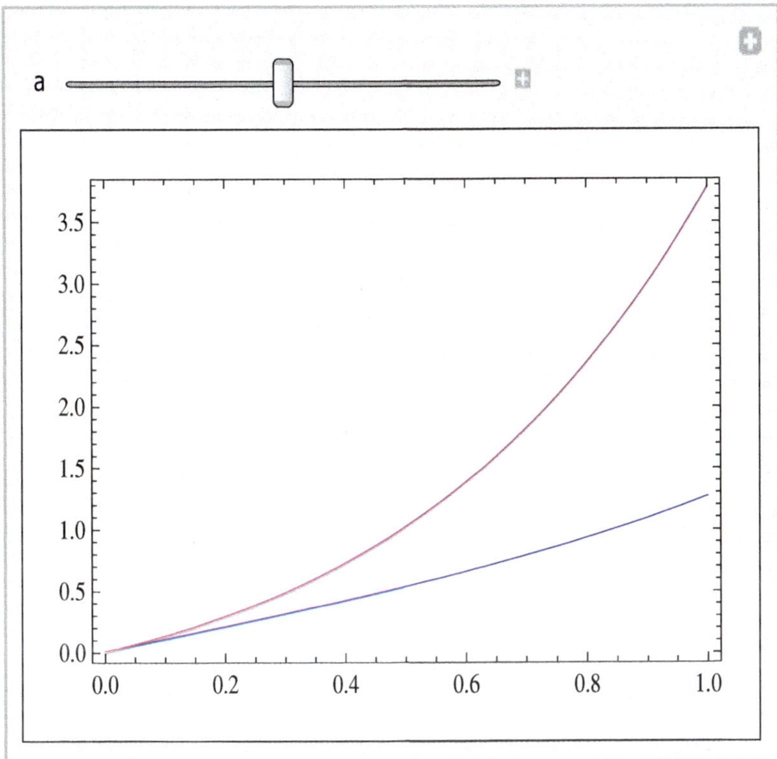

Fig. 4.1 Plots of the functions $f(t)$ and $y(t)$ from the Example 1

$$L[D^{\alpha}y(t)] = s^{\alpha}Y - s^{\alpha-1}y(0) = s^{\alpha}Y,$$

$$Y = \frac{1}{s^{\alpha+1}} + \frac{1}{s^{\alpha}}L\left[y^2\right].$$

For the decomposition

$$y(t) = \sum_{n=0}^{\infty} y_n(t),$$

we obtain

$$Y = \sum_{n=0}^{\infty} Y_n, \qquad y^2 = \sum_{n=0}^{\infty} A_n,$$

where A_n are the Adomian polynomials:

$$A_0 = y_0^2, \qquad A_1 = 2y_0y_1, \qquad A_2 = y_1^2 + 2y_0y_2, \qquad A_3 = 2y_1y_2 + 2y_0y_2, \ldots$$

$$\sum_{n=0}^{\infty} Y_n = \frac{1}{s^{\alpha+1}} + \frac{1}{s^{\alpha}} \sum_{n=0}^{\infty} A_n,$$

$$Y_0 = \frac{1}{s^{\alpha+1}} \quad \Rightarrow \quad y_0 = \frac{t^{\alpha}}{\Gamma(\alpha+1)},$$

$$A_0 = y_0^2 = \frac{t^{2\alpha}}{\Gamma^2(\alpha+1)},$$

$$Y_1 = \frac{1}{s^{\alpha}} L[A_0] \quad \Rightarrow \qquad Y_1 = \frac{1}{s^{\alpha}} L\left[\frac{t^{2\alpha}}{\Gamma^2(\alpha+1)}\right],$$

$$Y_1 = \frac{1}{s^{3\alpha+1}} \frac{\Gamma(2\alpha+1)}{\Gamma^2(\alpha+1)} \qquad \Rightarrow \ y_1 = L^{-1} Y_1,$$

$$y_1 = \frac{\Gamma(2\alpha+1)}{\Gamma^2(\alpha+1)} \frac{t^{3\alpha}}{\Gamma(3\alpha+1)},$$

$$A_1 = 2 y_0 y_1,$$

$$Y_2 = \frac{1}{s^{\alpha}} L[A_1],$$

$$Y_2 = 2 \frac{\Gamma(2\alpha+1)}{\Gamma^3(\alpha+1)} \frac{\Gamma(4\alpha+1)}{\Gamma(3\alpha+1)} \frac{1}{s^{5\alpha+1}},$$

$$y_2 = L^{-1}[Y_2],$$

$$y_2 = 2 \frac{\Gamma(2\alpha+1)}{\Gamma^3(\alpha+1)} \frac{\Gamma(4\alpha+1)}{\Gamma(3\alpha+1)\Gamma(5\alpha+1)} t^{5\alpha},$$

$$y(t) = y_0(t) + y_1(t) + y_2(t) + \dots$$

$$y(t) = \frac{t^{\alpha}}{\Gamma(\alpha+1)} + \frac{\Gamma(2\alpha+1)}{\Gamma^2(\alpha+1)} \frac{t^{3\alpha}}{\Gamma(3\alpha+1)} + 2 \frac{\Gamma(2\alpha+1)}{\Gamma^3(\alpha+1)} \frac{\Gamma(4\alpha+1)}{\Gamma(3\alpha+1)\Gamma(5\alpha+1)} t^{5\alpha} + \dots$$

You can also use the programs:

MAPLE

```
ec:=diff(y(t),t) = 1 + y(t))^2;
dsolve({ec,y(0) = 0},y(t),type = series);
```

MATHEMATICA

```
ec:=y'[t] == 1 + y[t]*y[t];
sol=DSolve[{ec,y[0]==0},y,t];
Series[y[t]/.sol,{t,0,10}]
```

Finally, we have:

$$y(t) = t + \frac{t^3}{3} + \frac{2t^5}{15} + \dots$$

Example 3 Solve the Ghelfand's[6] FDE:

$$D^{2\alpha} y(t) = 2e^{y(t)}, \qquad 0 < \alpha \leq 1,$$

where:

$$y(0) = y^{\alpha}(0) = 0,$$

using the Adomian decomposition method.

Solution To solve this problem we apply the LT:

$$L[D^{2\alpha} y(t)] = 2L\left[e^{y(t)}\right],$$

$$L[y(t)] = Y, \qquad y_0(t) = y(0) + \frac{y^{\alpha}(0)}{\Gamma(\alpha+1)} t^{\alpha} = 0,$$

$$L[D^{2\alpha} y(t)] = s^{2\alpha} Y - s^{2\alpha-1} y(0) = s^{\alpha} Y,$$

$$L\left[e^{y(t)}\right] = L\left[\sum_{n=0}^{\infty} A_n\right].$$

For the decomposition

$$y(t) = \sum_{n=0}^{\infty} y_n(t),$$

we obtain

$$Y = \sum_{n=0}^{\infty} Y_n, \qquad y^2 = \sum_{n=0}^{\infty} A_n,$$

[6] I.M. Ghelfand (1913–2009).

where A_n are the Adomian polynomials:

$$A_0 = e^{y_0}, \qquad A_1 = y_1 e^{y_0}, \ldots$$

$$\sum_{n=0}^{\infty} Y_n = \frac{2}{s^{2\alpha}} L\left[\sum_{n=0}^{\infty} A_n\right],$$

$$y_0 = 0,$$

$$A_0 = e^{y_0} = 1,$$

$$Y_1 = \frac{1}{s^{2\alpha}} L[A_0] \qquad \Rightarrow \qquad Y_1 = \frac{2}{s^{2\alpha+1}},$$

$$y_1 = \frac{2t^{2\alpha}}{\Gamma(2\alpha+1)},$$

$$A_1 = y_1 e^{y_0},$$

$$Y_2 = \frac{4}{s^{4\alpha+1}},$$

$$y_2 = L^{-1}[Y_2],$$

$$y_2 = \frac{4t^{4\alpha}}{\Gamma(4\alpha+1)},$$

$$y(t) = y_0(t) + y_1(t) + y_2(t) + \ldots$$

$$y(t) = 0 + \frac{2t^{2\alpha}}{\Gamma(2\alpha+1)} + \frac{4t^{4\alpha}}{\Gamma(4\alpha+1)} + \ldots$$

It can be used also the program:

MATHEMATICA

```
Clear["`*"]
Manipulate[
 f[t_] := t^2 + t^4/6;
 y[t_] := 2*t^(2*a)/Gamma[2*a + 1] + $*t^(4*a)/Gamma
 [4*a + 1]; Plot[{f[t], y[t]}, {t, 0, 1},
 ImageSize -> 300, Frame -> True], {{a, 1/2}, 0, 1}]
```

Figure 4.2 shows the plots of the functions $f(t)$ and $y(t)$.

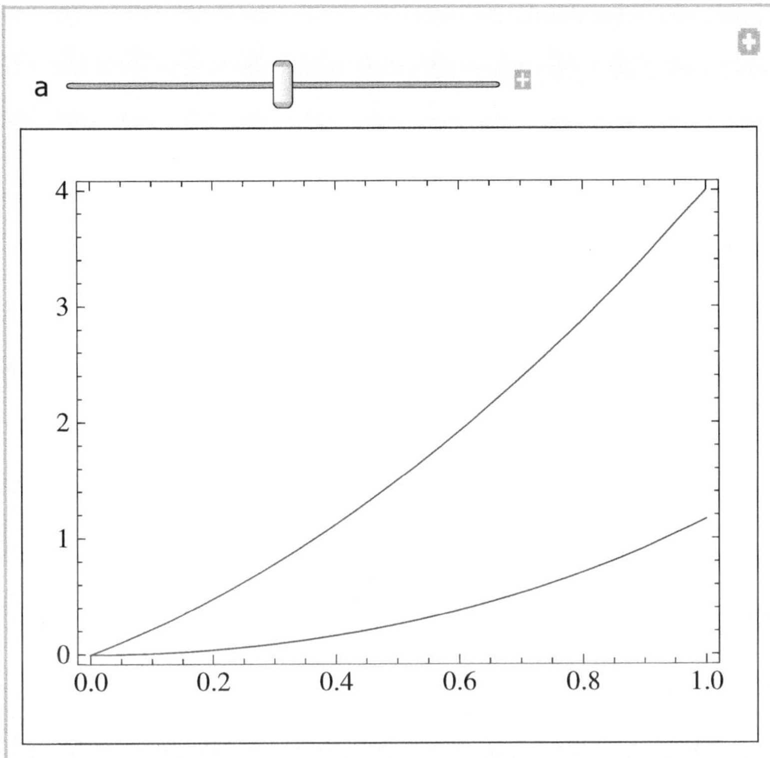

Fig. 4.2 Plots of the functions $f(t)$ and $y(t)$ from the Example 3

4.3.3 Perturbation Method

Here, we will extend the perturbation method for the case of FDE with the aid of some examples.

Example 1 Find the solution of the FDE:

$$D^\alpha y(t) = 1 + y^2(t), \qquad 0 < \alpha \leq 1,$$

for the initial condition:

$$y(0) = 0,$$

using the small parameter (perturbation) method, $0 < \epsilon \ll 1$.

Solution We consider a solution of form:

$$y(t) = y_0(t) + \epsilon y_1(t) + \epsilon^2 y_2(t) + \epsilon^3 y_3(t) + \dots,$$

which, replaced in the equation, gives:

$$D^\alpha y_0(t) = 1, \qquad y_0(0) = 0,$$

$$D^\alpha y_1(t) = y_0^2, \qquad y_1(0) = 0,$$

$$D^\alpha y_2(t) = 2y_0 y_1, \qquad y_2(0) = 0,$$

$$\dots$$

We apply the LT:

$$L[D^\alpha y_0(t)] = L[1], \qquad y_0(0) = 0,$$

$$s^\alpha Y_0 - y_0(0) = \frac{1}{s}, \qquad Y_0 = \frac{1}{s^{\alpha+1}} \qquad \Rightarrow \qquad y_0(t) = \frac{t^\alpha}{\Gamma(\alpha+1)},$$

$$L[D^\alpha y_1)] = L[y_0^2], \qquad y_1(0) = 0,$$

$$s^\alpha Y_1 = \frac{1}{\Gamma^2(\alpha+1)} \frac{\Gamma(2\alpha+1)}{s^{2\alpha+1}}, \qquad Y_1 = \frac{\Gamma(2\alpha+1)}{\Gamma^2(\alpha+1)} \frac{1}{s^{3\alpha+1}},$$

and by inverse LT, we obtain:

$$y_1(t) = \frac{\Gamma(2\alpha+1)}{\Gamma^2(\alpha+1)} \frac{t^{3\alpha}}{\Gamma(3\alpha+1)}.$$

It results also:

$$L[D^\alpha y_2] = L[2y_0 y_1], \qquad y_2(0) = 0,$$

$$s^\alpha Y_2 = 2\left[\frac{t^\alpha}{\Gamma(\alpha+1)}\right] L\left[\frac{\Gamma(2\alpha+1)}{\Gamma^2(\alpha+1)} \frac{t^{3\alpha}}{\Gamma(3\alpha+1)}\right],$$

$$s^\alpha Y_2 = 2\frac{\Gamma(2\alpha+1)}{\Gamma^3(\alpha+1)\Gamma(3\alpha+1)} L\left[t^{4\alpha}\right],$$

$$Y_2 = 2\frac{\Gamma(2\alpha+1)}{\Gamma^3(\alpha+1)\Gamma(3\alpha+1)} \frac{\Gamma(4\alpha+1)}{s^{5\alpha+1}},$$

$$y_2 = 2\frac{\Gamma(2\alpha+1)}{\Gamma^3(\alpha+1)\Gamma(3\alpha+1)} \frac{\Gamma(4\alpha+1)}{\Gamma(5\alpha+1)} t^{5\alpha},$$

The solution for this example:

$$y(t) = y_0(t) + y_1(t) + y_2(t) + \dots,$$

will be:

$$y(t) = \frac{t^\alpha}{\Gamma(\alpha+1)} + \frac{\Gamma(2\alpha+1)}{\Gamma^2(\alpha+1)}\frac{t^{3\alpha}}{\Gamma(3\alpha+1)} +$$

$$+2\frac{\Gamma(2\alpha+1)}{\Gamma^3(\alpha+1)\Gamma(3\alpha+1)}\frac{\Gamma(4\alpha+1)t^{5\alpha}}{\Gamma(5\alpha+1)} + \ldots$$

The plot of $f(t)$ and $y(t)$ is done with the following program and presented in Fig. 4.3.

MATHEMATICA

```
Clear["`*"]
Manipulate[
 f[t_] := t + t^3/3 + 2/15*t^5;
 y[t_] :=
  t^a/Gamma[a + 1] +
```

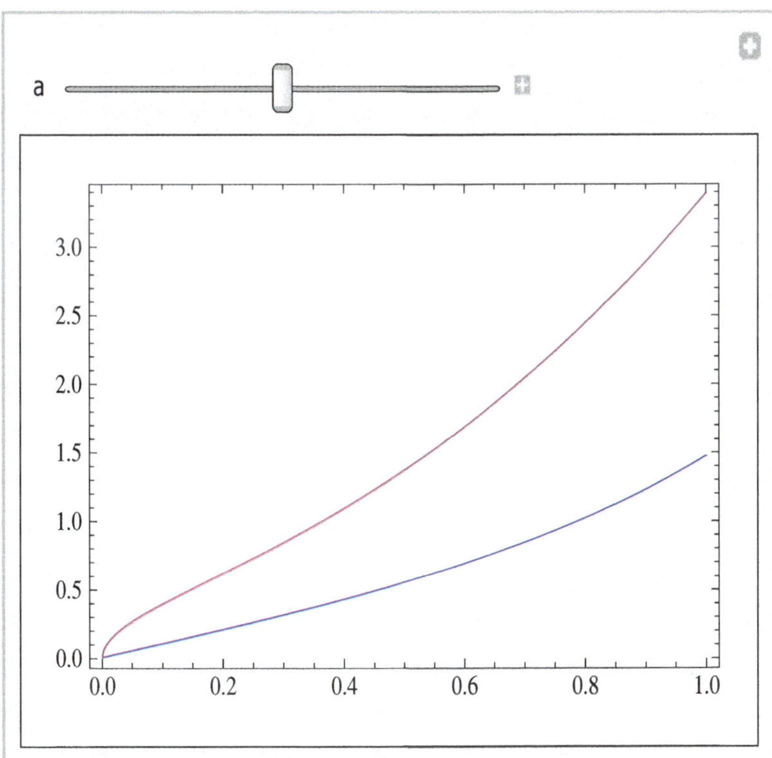

Fig. 4.3 Plots of the functions $f(t)$ and $y(t)$ from the above Example 1

```
    Gamma[2*a + 1]/(Gamma[a + 1])^2*t^(3*a)/Gamma
    [3*a + 1] + 2 Gamma[2*a + 1]/((Gamma[a + 1])^3*Gamma
    [3*a + 1])* Gamma[4*a + 1]/Gamma[5*a + 1]*t^(5*a);
  Plot[{f[t], y[t]}, {t, 0, 1}, ImageSize -> 300,
    Frame -> True], {{a, 1/2}, 0, 1}]
```

Example 2 Find the solution of the FDE:

$$D^{\alpha} y(t) = \frac{-y(t)}{1+\epsilon}, \qquad 0 < \alpha \leq 1,$$

for the initial condition:

$$y(0) = \cos(\epsilon), \qquad 0 < \epsilon \ll 1,$$

using the small parameter (perturbation) method.

Solution We consider a solution of form:

$$y(t) = y_0(t) + \epsilon y_1(t) + \epsilon^2 y_2(t) + \epsilon^3 y_3(t) + \ldots,$$

$$\frac{1}{1+\epsilon} = 1 - \epsilon + \epsilon^2 - \epsilon^3 + \ldots$$

$$\cos(\epsilon) = 1 - \frac{1}{2!}\epsilon^2 + \frac{1}{4!}\epsilon^4 + \ldots,$$

which, replaced in the equation, gives:

$$D^{\alpha}(y_0 + \epsilon y_1 + \epsilon^2 y_2 + \ldots) = -(y_0 + \epsilon y_1 + \epsilon^2 y_2 + \ldots)(1 - \epsilon + \epsilon^2 + \ldots)$$

$$D^{\alpha} y_0 = -y_0, \qquad y_0(0) = 1,$$

$$D^{\alpha} y_1 = y_0 - y_1, \qquad y_1(0) = 0,$$

$$D^{\alpha} y_2 = -y_2 + y_1 + y_0, \qquad y_2(0) = -\frac{1}{2},$$

$$\ldots$$

We apply the LT:

$$L[D^{\alpha} y_0(t)] = -L[y_0], \qquad y_0(0) = 1,$$

$$s^{\alpha} Y_0 - y_0(0) s^{\alpha-1} = -Y_0, \qquad Y_0 = \frac{s^{\alpha-1}}{s^{\alpha}+1}$$

$$L[D^\alpha y_1] = L[y_0] - L[y_1], \qquad y_1(0) = 0,$$

$$L[D^\alpha y_2] = -L[y_2] + L[y_1] + L[y_0], \qquad y_2(0) = -\frac{1}{2},$$

$$(s^\alpha + 1)Y_1 = Y_0, \qquad Y_1 = \frac{s^\alpha}{(s^\alpha + 1)^2},$$

$$Y_2 = \frac{s^{\alpha-1}}{(s^\alpha + 1)^3} + \frac{s^{\alpha-1}}{(s^\alpha + 1)^2} - \frac{1}{2}\frac{s^\alpha}{s^\alpha + 1},$$

$$y_0(t) = 1 - \frac{t^\alpha}{\Gamma(\alpha + 1)} + \frac{t^{2\alpha}}{\Gamma(2\alpha + 1)} - \frac{t^{3\alpha}}{\Gamma(3\alpha + 1)} + \dots,$$

$$Y_1 = \frac{1}{s^{\alpha+1}} \frac{1}{\left(1 + \frac{1}{s^\alpha}\right)^2},$$

$$Y_1 = \frac{1}{s^{\alpha+1}} - 2\frac{1}{s^{2\alpha+1}} + 3\frac{1}{s^{4\alpha+1}} + \dots,$$

and by inverse LT, we obtain:

$$y_1(t) = \frac{t^\alpha}{\Gamma(\alpha + 1)} - 2\frac{t^{2\alpha}}{\Gamma(2\alpha + 1)} + 3\frac{t^{4\alpha}}{\Gamma(4\alpha + 1)} + \dots$$

But:

$$\frac{s^{\alpha-1}}{s^{3\alpha}\left(1 + \frac{1}{s^\alpha}\right)^3} = \frac{1}{s^{2\alpha+1}}\frac{1}{2}\left[2 - 2 \cdot 3\frac{1}{s^\alpha} + 4 \cdot 3\frac{1}{s^{2\alpha}} + \dots\right].$$

It results also:

$$\begin{aligned} y_2(t) = {} & \frac{1}{2}\left[2\frac{t^{2\alpha}}{\Gamma(2\alpha + 1)} - 2 \cdot 3\frac{t^{3\alpha}}{\Gamma(3\alpha + 1)} + \dots\right] \\ & + \left[\frac{t^\alpha}{\Gamma(\alpha + 1)} - 2\frac{t^{2\alpha}}{\Gamma(2\alpha + 1)} + 3\frac{t^{4\alpha}}{\Gamma(4\alpha + 1)} + \dots\right] \\ & - \frac{1}{2}\left[1 - \frac{t^\alpha}{\Gamma(\alpha + 1)} + \frac{t^{2\alpha}}{\Gamma(2\alpha + 1)} - \frac{t^{3\alpha}}{\Gamma(3\alpha + 1)} + \dots\right]. \end{aligned}$$

4.4 Fractional Systems of Differential Equations

4.4.1 Linear Systems

Examples Solve the system of FDE:

$$\begin{cases} D^{\alpha}x(t) = D^{\beta}y(t) + 1, & x(0) = 1, \quad 0 < \alpha \leq 1 \\ D^{\beta}y(t) = 2D^{\alpha}x(t) - 1, & y(0) = 1, \quad 0 < \beta \leq 1. \end{cases},$$

Solution We apply the LT method:

$$L[x(t)] = X, \quad L[y(t)] = Y,$$

$$L[D^{\alpha}x(t)] = s^{\alpha}X - s^{\alpha-1}x(0) = s^{\alpha}X - s^{\alpha-1},$$

$$L[D^{\beta}y(t)] = s^{\beta}Y - s^{\beta-1}y(0) = s^{\beta}Y - s^{\beta-1}.$$

We obtain the system

$$\begin{cases} X = \dfrac{1}{s}, \\ Y = \dfrac{1}{s} - \dfrac{1}{s^{\beta+!}}, \end{cases}$$

with the solution:

$$\begin{cases} x(t) = 1, \\ y(t) = 1 - \dfrac{t^{\beta}}{\Gamma(\beta+1)}. \end{cases}$$

4.4.2 Nonlinear Systems

(A) Method of Successive Approximations

For the system of FDE:

$$\begin{cases} D^{\alpha}x(t) = f(t, y(t)), & x(0) = x_0, \\ D^{\alpha}y(t) = g(t, x(t)), & y(0) = y_0, \end{cases}$$

we can use the following successive approximations [4]:

$$\begin{cases} x_n(t) = x_0 + \dfrac{1}{\Gamma(\alpha)} \displaystyle\int_0^t f(u, y_{n-1}(u))(t-u)^{\alpha-1} du, \\ y_n(t) = y_0 + \dfrac{1}{\Gamma(\alpha)} \displaystyle\int_0^t g(u, x_{n-1}(u))(t-u)^{\alpha-1} du. \end{cases}$$

Example We apply the successive approximation method for the system of FDE with initial conditions:

$$\begin{cases} D^\alpha x(t) = 3.5 y(t)(1 - y(t)), \qquad x(0) = 0.2, \\ D^\alpha y(t) = 4 x(t)(1 - x(t)), \qquad y(0) = 0.2. \end{cases}$$

We have:

$$\begin{cases} x_n(t) = 0.2 + \dfrac{3.5}{\Gamma(\alpha)} \displaystyle\int_0^t \left(y_{n-1}(u) - y_{n-1}^2(u)\right)(t-u)^{\alpha-1} du, \\ y_n(t) = 0.2 + \dfrac{4}{\Gamma(\alpha)} \displaystyle\int_0^t \left(x_{n-1}(u) - x_{n-1}^2(u)\right)(t-u)^{\alpha-1} du, \end{cases}$$

$$\begin{cases} x_1 = 0.2 + \dfrac{3.5}{\Gamma(\alpha)} \displaystyle\int_0^t \left(0.2 - (0.2)^2\right)(t-u)^{\alpha-1} du, \\ y_1 = 0.2 + \dfrac{4}{\Gamma(\alpha)} \displaystyle\int_0^t \left(0.2 - (0.2)^2\right)(t-u)^{\alpha-1} du. \end{cases}$$

Using the theorem regarding the product of convolution, we obtain the following three iterations:

$$\begin{cases} x_1(t) = 0.2 + 0.56 \dfrac{t^\alpha}{\Gamma(\alpha+1)}, \\ y_1(t) = 0.2 + 0.56 \dfrac{t^\alpha}{\Gamma(\alpha+1)}. \end{cases}$$

$$\begin{cases} x_2(t) = 0.2 + \dfrac{3.5}{\Gamma(\alpha)} \displaystyle\int_0^t \left(y_1(u) - y_2^1(u)\right)(t-u)^{\alpha-1} du, \\ y_2(t) = 0.2 + \dfrac{4}{\Gamma(\alpha)} \displaystyle\int_0^t \left(x_1(u) - x_2^1(u)\right)(t-u)^{\alpha-1} du, \end{cases}$$

$$\begin{cases} x_2(t) = 0.2 \\ +0.56\dfrac{t^{\alpha}}{\Gamma(\alpha+1)} + 1.344\dfrac{t^{2\alpha}}{\Gamma(2\alpha+1)} - 0.808\dfrac{\Gamma(2\alpha+1)}{\Gamma^2(\alpha+1)}\dfrac{t^{3\alpha}}{\Gamma(3\alpha+1)}, \\ y_2(t) = 0.2 \\ +0.64\dfrac{t^{\alpha}}{\Gamma(\alpha+1)} + 1.344\dfrac{t^{2\alpha}}{\Gamma(2\alpha+1)} - 0.7077\dfrac{\Gamma(2\alpha+1)}{\Gamma^2(\alpha+1)}\dfrac{t^{3\alpha}}{\Gamma(3\alpha+1)}, \end{cases}$$

$$\begin{cases} x_3(t) = 0.2 + \dfrac{3.5}{\Gamma(\alpha)}\displaystyle\int_0^t \left(y_2(u) - y_2^2(u)\right)(t-u)^{\alpha-1}du \\ y_3(t) = 0.2 + \dfrac{4}{\Gamma(\alpha)}\displaystyle\int_0^t \left(x_2(u) - x_2^2(u)\right)(t-u)^{\alpha-1}du \end{cases}$$

$$\cdots$$

For $\alpha = 0.9$ we can apply the Maple and Mathematica programs:

MAPLE

```
> restart;
> with(inttrans):
> Digits:=5:
> x:=array(0..10):
> y:=array(0..10):
> x[0]:=0.2:
> y[0]:=0.2:
> for k from 1 to 5 do
> x[k]:=evalf(0.2+3.5*invlaplace(1/s^0.9*laplace
                                  (y[k-1]-(y[k-1])^2,
                                  t,s),s,t));
> y[k]:=evalf(0.2+4*invlaplace(1/s^0.9*laplace
                                (x[k-1]-(y[k-1])^2,
                                t,s),s,t));
> od:
> for k from 0 to 4 do
> print([x[k],y[k]]) od:
 x                         y
 0.2                        0.2
 0.2 + 0.58230 t^(9/10)     0.2 + 0.66548t^(9/10)
 0.2 + 0.58230 t^(9/10) + 0.80171 t^(9/5) - 0.62304
          t^(27/10),
 0.2 + 0.66548 t^(9/10) + 0.72536 t^(9/5) - 0.71204
           t^(27/10)
                 . . . . .
```

MATHEMATICA

```
Clear["`*"]
Array[x, 10]
Array[y, 10]
For[{n = 0, x[0] = 0.2, y[0] = 0.2}, n < 4,
 n++, {x[n + 1] =
   0.2 + 3.5 InverseLaplaceTransform[
   1/s^0.9 LaplaceTransform[y[n] - (y[n])^2, t, s],
    s, t]// FullSimplify,
y[n + 1] =
 0.2 + 4 InverseLaplaceTransform[
 1/s^0.9 LaplaceTransform[(x[n] - (x[n])^2), t, s],
     s, t]//
 FullSimplify, Print["x=", x[n], " , ", "y= ", y[n]]}]
  x = 0.2,                    y=0.2
  x = 0.2+0.582262 t^0.9 , y= 0.2+0.665443 t^0.9
```

(B) Method of Laplace's Transform

We will illustrate this method on the function:

$$F(y) = y - y^2.$$

First, we will decompose F in terms of Adomian's polynomials

$$F = \sum_{n=0}^{\infty} A_n,$$

where $A_0 = y_0 - y_0^2$ and

$$\phi_1(\lambda) = (y_0 + \lambda y_1) - (y_0 + \lambda y_1)^2,$$

$$\phi_1' = y_1 - 2y_1(y_0 + \lambda y_1),$$

$$A_1 = \frac{1}{1!}\phi_1(0),$$

from which we obtain: $A_1 = y_1 - 2y_0y_1$.

In the case of next step we have:

$$\phi_2(\lambda) = (y_1 + 2\lambda y_2) - 2(y_0 + \lambda y_1)(y_1 + 2\lambda y_2),$$

$$\phi_2' = 2y_2 - 2y_1(y_1 + 2\lambda y_2) - 2y_2(y_0 + \lambda y_1),$$

$$A_2 = \frac{1}{2!}\phi_2'(0),$$

or finally: $A_2 = y_2 - y_1^2 - y_2y_0$.

We have also:

$$\phi_3(\lambda) = (y_2 + 3\lambda y_3) - (y_1 + 2\lambda y_2)^2 - (y_2 + 3\lambda y_3)(y_0 + \lambda y_1),$$

$$A_3 = \frac{1}{3!}\phi_3'(0),$$

$$\dots$$

We consider now a system described by the equations [8]:

$$\begin{cases} L[D^\alpha x(t)] = 3.5L[y(t)(1 - y(t))], & x(0) = 0.2, \\ L[D^\alpha y(t)] = 4L[x(t)(1 - x(t))], & y(0) = 0.2, \end{cases}$$

with initial conditions and we apply the LT to this system. We have:

$$L[x(t)] = X, \qquad L[y(t)] = Y,$$

$$L[D^\alpha x(t)] = s^\alpha - x(0)s^{\alpha-1},$$

$$L[D^\alpha y(t)] = s^\alpha - y(0)s^{\alpha-1},$$

We consider the solutions:

$$X = \sum_{n=0}^{\infty} X_n, \qquad Y = \sum_{n=0}^{\infty} Y_n.$$

After calculations we have:

$$L\left[x(t)(1 - x(t))\right] = L\left[\sum_{n=0}^{\infty} A_n\right], \qquad L\left[y(t)(1 - y(t))\right] = L\left[\sum_{n=0}^{\infty} B_n\right],$$

where A_n and B_n are Adomian's polynomials.

$$\sum_{n=0}^{\infty} X_n = \frac{0.2}{s} + \frac{3.5}{s^\alpha} L\left[\sum_{n=0}^{\infty} B_n\right]$$

$$\sum_{n=0}^{\infty} Y_n = \frac{0.2}{s} + \frac{4}{s^\alpha} L\left[\sum_{n=0}^{\infty} A_n\right]$$

$$X_0 = \frac{0.2}{s} \quad \Rightarrow \quad x_0 = 0.2$$

$$Y_0 = \frac{0.2}{s} \quad \Rightarrow \quad x_0 = 0.2$$

$$\begin{cases} X_1 = \dfrac{3.5}{s^\alpha} L[B_0] = \dfrac{0.56}{s^{\alpha+1}}, & \Rightarrow \quad x_1(t) = 0.56 \dfrac{t^\alpha}{\Gamma(\alpha+1)} \\ Y_1 = \dfrac{4}{s^\alpha} L[A_0] = \dfrac{0.64}{s^{\alpha+1}}, & \Rightarrow \quad y_1(t) = 0.64 \dfrac{t^\alpha}{\Gamma(\alpha+1)} \end{cases}$$

$$\begin{aligned} B_1 &= y_1 - 2y_0 y_1 = y_1(1 - 2y_0) = 0.64 \frac{t^\alpha}{\Gamma(\alpha+1)}(1 - 2 \cdot 0.2) \\ &= 0.64 \cdot 0.6 \cdot \frac{t^\alpha}{\Gamma(\alpha+1)} = 0.384 \frac{t^\alpha}{\Gamma(\alpha+1)} \end{aligned}$$

$$\begin{aligned} A_1 &= x_1 - 2x_0 x_1 = x_1(1 - 2x_0) = 0.56 \frac{t^\alpha}{\Gamma(\alpha+1)}(1 - 2 \cdot 0.2) \\ &= 0.56 \cdot 0.6 \cdot \frac{t^\alpha}{\Gamma(\alpha+1)} = 0.336 \frac{t^\alpha}{\Gamma(\alpha+1)} \end{aligned}$$

$$X_2 = \frac{3.5}{s^\alpha} L[B_1] = \frac{1.344}{s^{2\alpha}} \quad \Rightarrow x_2(t) = 1.344 \frac{t^{2\alpha}}{\Gamma(2\alpha+1)}$$

$$Y_2 = \frac{4}{s^\alpha} L[A_1] = \frac{1.344}{s^{2\alpha}} \quad \Rightarrow y_2(t) = 1.344 \frac{t^{2\alpha}}{\Gamma(2\alpha+1)}$$

Finally, the solution is:

$$\begin{cases} x(t) = x_0(t) + x_1(t) + x_2(t) + \ldots = 0.2 + 0.56 \dfrac{t^\alpha}{\Gamma(\alpha+1)} + 1.344 \dfrac{t^{2\alpha}}{\Gamma(2\alpha+1)} + \ldots \\ y(t) = y_0(t) + y_1(t) + y_2(t) + \ldots = 0.2 + 0.64 \dfrac{t^\alpha}{\Gamma(\alpha+1)} + 1.344 \dfrac{t^{2\alpha}}{\Gamma(2\alpha+1)} + \ldots \end{cases}$$

References

1. Adomian, G. (1988). A review of the decomposition method in applied mathematics. *Journal of Mathematical Analysis and Applications, 135*(2), 501–544.
2. Adomian, G. (1994). *Solving frontier problems of physics: The decomposition method. Fundamental theories of physics*. Dordrecht: Springer.
3. Cherruault, Y. (1989). Convergence of Adomian's method. *Kybernetes, 18*(2), 31–38.

4. El'sgol'ts, L. E., & Norkin, S. B. (1973). *Introduction to the theory of differential equations with deviating arguments. Mathematics in science and engineering*. New York: Academic Press.
5. Kazem, S. (2013). Exact solution of some linear differential equations by Laplace transform. *International Journal of Nonlinear Science, 16*, 3–11.
6. Khan, M., Hussain, M., Jafari, H., & Khan, Y. (2010). Application of Laplace decomposition method to solve nonlinear coupled partial differential equations. *World Applied Sciences Journal, 9*, 13–19.
7. Khelifa, S., & Cherruault, Y. (2000). New results for the Adomian method. *Kybernetes, 29*, 332–354.
8. Milici, C., & Drăgănescu, G. (2014). *A method for solve the nonlinear fractional differential equations*. Saarbrücken: Lambert Academic Publishing.
9. Natanson, I. P. (1950). *Teoria funcţii vescestvennoi peremennoi*. Gosudarstvennoe izdatelstvo tehniko-teoreticekoi literaturi, Moscva.
10. Pinkus, A. (2000). Weierstrass and approximation theory. *Journal of Approximation Theory, 107*, 1–66.
11. O'Shaughnessy, L. (1918). Problem 433. *The American Mathematical Monthly, 25*, 172.
12. Weilbeer, M. (2005). *Efficient numerical methods for fractional differential equations and their analytical background*. PhD thesis, Facultät für Mathematik und Informatik, Technischen Univerisität Braunschweig, Braunschweig.

Chapter 5
Generalized Systems

This chapter addresses the generalization of classical models and systems in the perspective of FC. The following sections study the Cornu, Emden, Hermite, Legendre, and Bessel fractional systems.

5.1 Cornu Fractional System

5.1.1 Cos and Sin Fractional of Type Fresnel

We define fractional Cos, and Sin of order $0 < q \leq 1$, as:

$$C^q(t^2) = \frac{1}{\Gamma(q)} \int_0^t (t-u)^{q-1} \cos u^2 du,$$

$$S^q(t^2) = \frac{1}{\Gamma(q)} \int_0^t (t-u)^{q-1} \sin u^2 du.$$

These functions can be represented in Mathematica with the aid of the following program for $q = 1/2$ (see Fig. 5.1).

MATHEMATICA

```
Clear["`*"]
q = 1/2;
u[t_] = 1/Gamma[q]*Integrate[(t-s)^(q-1)*Cos[s^2],
  {s, 0, t}];
```


C. Milici et al., *Introduction to Fractional Differential Equations*, Nonlinear Systems and Complexity 25, https://doi.org/10.1007/978-3-030-00895-6_5

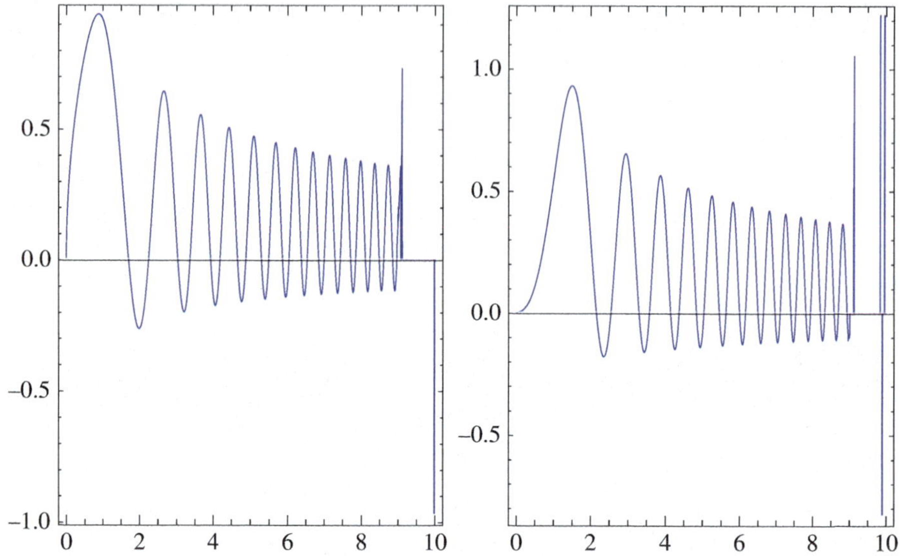

Fig. 5.1 The graph of fractional cosine (left) and sine (right) for $q = 1/2$

```
p = Plot[{u[t]}, {t, 0, 10}, PlotLabel -> "Cos
  fractional",
  ImageSize -> 200, Frame -> True];
v[t_] = 1/Gamma[q]*Integrate[(t-s)^(q-1)*Sin[s^2],
  {s, 0, t}];
  q = Plot[{v[t]}, {t, 0, 10}, PlotLabel -> "Sin
  fractional",
  ImageSize -> 200, Frame -> True];
  Show[GraphicsArray[{p, q}]]
```

5.1.2 Cornu Fractional System and Curve

We will introduce the following Cornu[1] fractional differential system with initial conditions:

$$\begin{cases} D^q x(t) = \cos(t^2), & x(0) = 0, \\ D^q y(t) = \sin(t^2), & y(0) = 0. \end{cases}$$

[1]M.A. Cornu (1841–1902).

Solution We will solve this system using the LT. We have:

$$\begin{cases} L\left[x^q(t)\right] = L\left[\cos(t^2)\right] = L\left[1 - \frac{t^4}{2!} + \frac{t^8}{4!} - \frac{t^{12}}{6!} + \ldots\right], \\ L\left[y^q(t)\right] = L\left[\sin(t^2)\right] = L\left[t^2 - \frac{t^6}{3!} + \frac{t^{10}}{5!} - \frac{t^{14}}{7!} + \ldots\right], \end{cases}$$

$$\begin{cases} s^q X = \frac{1}{s} - \frac{4!}{2!s^5} + \frac{8!}{4!s^9} - \frac{12!}{6!s^{13}} + \ldots, \\ s^q Y = \frac{2!}{s^3} - \frac{6!}{3!s^7} + \frac{10!}{5!s^{11}} - \frac{14!}{7!s^{15}} + \ldots, \end{cases}$$

$$\begin{cases} X = \frac{1}{s^{q+1}} - \frac{4!}{2!s^{q+5}} + \frac{8!}{4!s^{q+9}} - \frac{12!}{6!s^{q+13}} + \ldots, \\ Y = \frac{2!}{s^{q+3}} - \frac{6!}{3!s^{q+7}} + \frac{10!}{5!s^{q+11}} - \frac{14!}{7!s^{q+15}} + \ldots \end{cases}$$

By inverse LT we obtain the solution $\{x(t), y(t)\}$:

$$\begin{cases} x(t) = \frac{t^q}{\Gamma(q+1)} - \frac{4!}{2!}\frac{t^{q+4}}{\Gamma(q+5)} + \frac{8!}{4!}\frac{t^{q+8}}{\Gamma(q+9)} - \frac{12!}{6!}\frac{t^{q+12}}{\Gamma(q+13)} + \ldots \\ y(t) = \frac{2t^{q+2}}{\Gamma(q+3)} - \frac{6!}{3!}\frac{t^{q+6}}{\Gamma(q+7)} + \frac{10!}{5!}\frac{t^{q+10}}{\Gamma(q+11)} - \frac{14!}{7!}\frac{t^{q+14}}{\Gamma(q+15)} + \ldots \end{cases}$$

The graph of this curve can be written in Maple and Mathematica as:

MAPLE

```
restart;a:=1/2:
> x(t):=1/GAMMA(a)*int((t-u)^(a-1)*cos(u^2),u=0..t):
> y(t):=1/GAMMA(a)*int((t-u)^(a-1)*sin(u^2),u=0..t):
> plot([x(t),y(t),t=0..10],color=black,
  scaling=constrained);
```

MATHEMATICA

```
Clear["`*"]
q = 1/2;
u[t_] = 1/Gamma[q]*Integrate[(t - s)^(q - 1)*Cos[s^2],
                                              {s, 0, t}];
v[t_] = 1/Gamma[q]*Integrate[(t-s)^(q-1)*Sin[s^2],
 {s, 0, t}];
```

```
 ParametricPlot[{u[t], v[t]}, {t, 0, 10},
                             Frame -> True, ImageSize -> 300]

Clear["`*"]
Manipulate[
ParametricPlot[{1/Gamma[q]
  *Integrate[(t - u)^(q - 1)*Cos[u^2], {u, 0, t}],
  1/Gamma[q]*Integrate[(t - u)^(q - 1)*Sin[u^2],
  {u, 0, t}]}, {t, 0,  10}, Frame -> True,
  ImageSize -> 300], {{q, 3/2}, 0, 2}]
```

5.1.3 Cornu Generalized Curve/System

For $0 \le \alpha \le 1$ we can introduce a generalization of above curve and system, as:

$$\begin{cases} D^{\alpha}x(t) = \cos\dfrac{t^q}{\Gamma(q+1)}, & x(0) = 0 \\ D^{\alpha}y(t) = \sin\dfrac{t^q}{\Gamma(q+1)}, & y(0) = 0 \end{cases}$$

5.1.4 Cornu Fractional System in a Plane

This curve can be plotted (see Fig. 5.2) in a plane using Maple and Mathematica as:

MAPLE

```
> restart;
> with(plots):
> a:=1/2:
> x:=1/GAMMA(a)*int((t-u)^(a-1)*cos(u^2),u=0..t):
> y:=1/GAMMA(a)*int((t-u)^(a-1)*sin(u^2),u=0..t):
> spacecurve([x,y,3 - x - y,t = 0..4],color=black,
                                  scaling=constrained);
```

MATHEMATICA

```
Clear["`*"]
Manipulate[
 {x[t], y[t]} = {Integrate[Cos[u^2], {u, 0, t}],
   Integrate[Sin[u^2], {u, 0, t}]};
p = ({x, y, z} /. First@Solve[x/a + y/b + z/c - 1
                                      == 0, z]);
```

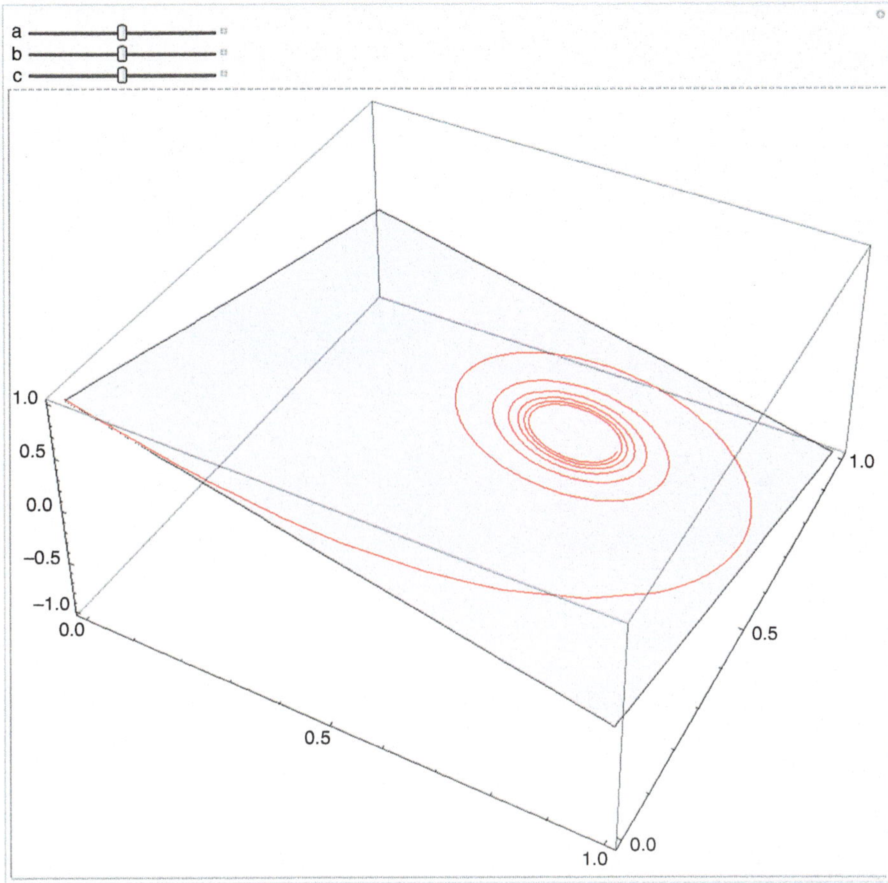

Fig. 5.2 Cornu fractional curve in plane

```
q = p /. {x -> x[t], y -> y[t]};
 Show[Plot3D[Evaluate[p[[3]]], {x, 0, 1},
                     {y, 0, 1}, Mesh->None,
   PlotStyle -> {Blue, Opacity[0.05]}],
  ParametricPlot3D[Evaluate[q], {t, 0, 2*Pi},
                          PlotStyle -> Blue],
  ImageSize->300, AspectRatio->1], {{a, 1}, 0, 2},
           {{b, 1}, 0, 2}, {{c, 1}, 0, 2}]
```

5.1.5 *Fractional Cornu Spiral on the Sphere*

Projection of the Cornu fractional curve on sphere (see Fig. 5.3) can be obtained in Maple and Mathematica as:

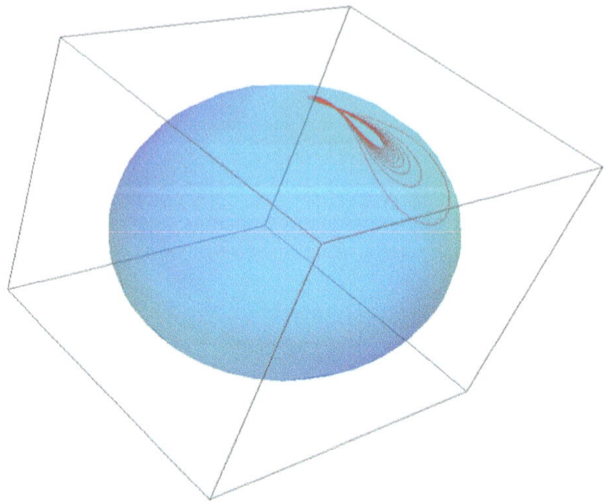

Fig. 5.3 Projection of the Cornu fractional curve on sphere

MAPLE

```
> restart;
> with(plots):
> a:=1/2:
> x:=1/GAMMA(a)*int((t-u)^(a-1)*cos(u^2),u=0..t):
> y:=1/GAMMA(a)*int((t-u)^(a-1)*sin(u^2),u=0..t):
> spacecurve([cos(x)*sin(y),sin(x)*sin(y),cos(y),
  t=0..4],
color=black,scaling=constrained);
```

MATHEMATICA

```
Clear["`*"]
p = ParametricPlot3D[{Cos[u]*Sin[v], Sin[u]*Sin[v],
  Cos[v]},
  {u, 0, 2*Pi}, {v, 0, Pi}, PlotStyle->{Cyan,
  Opacity[0.3]},
   Mesh -> None, Axes -> False, Boxed -> False];
   q = 1/2;
u[t_] = 1/Gamma[q]*Integrate[(t - s)^(q - 1)*Cos[s^2],
                                            {s, 0, t}];
v[t_] = 1/Gamma[q]*Integrate[(t - s)^(q - 1)*Sin[s^2],
                                            {s, 0, t}];
q1 =  ParametricPlot3D[{Cos[u[t]]*Sin[v[t]],
   Sin[u[t]]*Sin[v[t]], Cos[v[t]]}, {t, 0, 10},
              PlotStyle -> {Red, Opacity[0.3]},
```

```
Mesh -> None, Axes -> False, Boxed -> False];
Show[p, q1]
```

5.1.6 *Fractional Cornu Spiral on the Cone*

This projection (see Fig. 5.4) can be obtained with Mathematica as:

```
Clear["`*"]
x[u_, v_] := u*Cos[v];
y[u_, v_] := u*Sin[v];
z[u_, v_] := -u;
Manipulate[
u[t_] = 1/Gamma[q]*Integrate[(t - s)^(q - 1)*Cos[s^2],
                                              {s, 0, t}];
v[t_] = 1/Gamma[q]*Integrate[(t - s)^(q - 1)*Sin[s^2],
                                              {s, 0, t}];
p = ParametricPlot3D[{x[u, v], y[u, v], z[u, v]}, {
   u, -1, 1}, {v, 0, 2*Pi}, Mesh -> None, Axes
                                 -> False, Boxed -> False];
q1 = ParametricPlot3D[{x[u[t], v[t]], y[u[t], v[t]],
    z[u[t], v[t]]}, {t, 0, 10}, Mesh -> None,
    Axes -> False, Boxed -> False];
Show[p, q1], {{q, 1/2}, 0, 2}]
```

Fig. 5.4 Projection of the Cornu fractional spiral on the cone

5.2 Power Series

We will use here the power series method [3] to solve the FDE. We denote by:

$$\frac{1}{\Gamma(\alpha+1)}\int_0^t f(t)(dt)^\alpha = \frac{\alpha}{\Gamma(\alpha+1}\int_0^t (t-u)^{\alpha-1} f(u)du = I^\alpha f(t),$$

the FI of order α of $f(t)$. Then, the norm of order α on $L_{L^2_{[0,1]}}$ of $f(t)$ can be defined:

$$\|I^\alpha f(t)\|^\alpha_{L^2_{[0,1]}} = \left(\frac{1}{\Gamma(\alpha+1)}\int_0^1 f^2(t)(dt)^\alpha)\right)^{1/2}.$$

5.2.1 *The Müntz Theorem*

For details, regarding this theorem,[2] the reader can see [1, 7, 8].

We denote by $C[a,b]$ the set of all continuous functions on $[a,b]$.

Definition 4 For a function $f \in C[a,b]$ the norm is defined

$$\|f\| = \max_{t\in[a,b]} |f(t)|,$$

and for $f, g \in C[a,b]$ it can be defined the distance:

$$\|f-g\| = \max_{t\in[a,b]} |f(t)-g(t)|.$$

Definition 5 A sequence $f_n(t)$ of functions in $C[a,b]$ converges uniformly to a function $f(t)$ if $\lim\limits_{n\to\infty} \|f_n(t)-f(t)\| = 0$.

Definition 6 Let X be a metric space. If $A \subset X$ then

$$\overline{A} = A\bigcup\{\lim_n a_n : \forall n>0, \quad a_n \in A\},$$

is **dense** in A if:

$$\overline{A} = X.$$

Theorem 7 *The system*

$$\Pi(\Lambda) = span\{t^{\alpha_0}, t^{\alpha_1}, \ldots, t^{\alpha_n}, \ldots\}, \qquad \alpha_k \in \mathbb{R},$$

[2]H. Müntz (1884–1956).

where

$$\Lambda = \{0 = \alpha_0 < \alpha_1 < \alpha_2 < \ldots\},$$

is dense in $C[0, 1]$, *if:*

$$\sum_{n=1}^{\infty} \frac{1}{\alpha_n} = \infty.$$

Proof We will establish the expression of error in $L^2[0, 1]$. We have:

$$\|t^q - p(t)\|_{C[0,1]} < \|qt^{q-1} - p'(t)\|_{L^2[0,1]},$$

$$p(t) \in \Pi(\Lambda),$$

$$\begin{aligned} \mathrm{Er}(t^q, p(t))_{L^2[0,1]}) = I[C_1, C_2, \ldots, C_n] &= \int_0^1 [t^q - p(t)]^2 \, dt \\ &= \int_0^1 \left[t^q - \sum_{k=1}^{n} C_k t^{\alpha_k} \right]^2 dt \to \min, \end{aligned}$$

where $t \in [0, 1]$, $p(t) \in \text{span}\{1, t^{\alpha_1}, t^{\alpha_2}, \ldots, \}$, we have the functional:

$$I[C_1, C_2, \ldots, C_n] = \int_0^1 \left[t^q - \sum_{k=1}^{n} C_k t^{\alpha_k} \right]^2 dt \to \min.$$

which, by minimization, give us the following system of equations in $\{C_1, C_2, \ldots, C_n\}$:

$$\frac{\partial I[C_1, C_2, \ldots, C_n]}{\partial C_i} = 0, \quad i = 1, 2, \ldots n.$$

This system can be written in an explicit form as:

$$\begin{cases} C_1 \dfrac{1}{\alpha_1 + \alpha_1 + 1} + C_2 \dfrac{1}{\alpha_1 + \alpha_2 + 1} + \ldots + C_n \dfrac{1}{\alpha_1 + \alpha_n + 1} = \dfrac{1}{\alpha_1 + q + 1}, \\ C_1 \dfrac{1}{\alpha_2 + \alpha_1 + 1} + C_2 \dfrac{1}{\alpha_2 + \alpha_2 + 1} + \ldots + C_n \dfrac{1}{\alpha_2 + \alpha_n + 1} = \dfrac{1}{\alpha_2 + q + 1}, \\ \qquad \ldots \qquad \ldots \qquad \ldots \\ C_1 \dfrac{1}{\alpha_n + \alpha_1 + 1} + C_2 \dfrac{1}{\alpha_n + \alpha_2 + 1} + \ldots + C_n \dfrac{1}{\alpha_n + \alpha_n + 1} = \dfrac{1}{\alpha_n + q + 1}. \end{cases}$$

In the above system it appears the symmetric matrix:

$$H(\alpha_1, \alpha_2, \ldots, \alpha_n) = \begin{pmatrix} \frac{1}{\alpha_1+\alpha_1+1} & \frac{1}{\alpha_1+\alpha_2+1} & \cdots & \frac{1}{\alpha_1+\alpha_n+1} \\ \frac{1}{\alpha_2+\alpha_1+1} & \frac{1}{\alpha_2+\alpha_2+1} & \cdots & \frac{1}{\alpha_2+\alpha_n+1} \\ \cdots & \cdots & \cdots & \\ \frac{1}{\alpha_n+\alpha_1+1} & \frac{1}{\alpha_n+\alpha_2+1} & \cdots & \frac{1}{\alpha_n+\alpha_n+1} \end{pmatrix},$$

having the Gram[3] determinant:

$$G(\alpha_1, \alpha_2, \ldots, \alpha_n) = \det(H(\alpha_1, \alpha_2, \ldots, \alpha_n)),$$

where Determinant (det) denotes determinant.

It is well known that:

$$\mathrm{Er}(t^q, p(t))_{L^2[0,1]} = \sqrt{\frac{G(q, \alpha_1, \alpha_2, \ldots, \alpha_n)}{G(\alpha_1, \alpha_2, \ldots, \alpha_n)}},$$

$$\mathrm{Er}(t^q, p(t))_{L^2[0,1]} = \frac{1}{\sqrt{2q+1}} \prod_{k=1}^{n} \frac{|q-\alpha_k|}{q+\alpha_k+1}.$$

By recurrence, we can build the following set of functions:

$$Q_0(t) = t^q,$$

$$\ldots$$

$$Q_n(t) = (\alpha_n - q)t^{\alpha_n} \int_t^1 Q_{n-1}(u)u^{-(1+\alpha_n)}\, du.$$

After calculations, we obtain:

$$Q_1 = (\alpha_1 - q)t^{\alpha_1} \int_t^1 u^{-1-\alpha_1} u^q du = t^q - t^{\alpha_1},$$

$$Q_n(t) = t^q - \sum_{k=0}^{n-1} C_k t^{\alpha_k}.$$

[3] J.P. Gram (1850–1916).

The last function can be verified by induction

$$Q_n(t) = (\alpha_n - q)t^{\alpha_n} \int_t^1 u^{-(1+\alpha_n)} Q_{n-1}(u)\, du,$$

$$Q_n(t) = (\alpha_n - q)t^{\alpha_n} \int_t^1 u^{-(1+\alpha_n)} \left(u^q - \sum_{k=0}^{n-2} C_k u^{\alpha_k} \right) du,$$

$$Q_n(t) = t^q - t^{\alpha_n} + (\alpha_n - q) \sum_{k=0}^{n-2} \frac{C_k}{\alpha_n - \alpha_k} (t^{\alpha_k} - t^{\alpha_n}),$$

or, using the inequality

$$\alpha t^{\alpha}(1-t) < 1, \qquad \text{where: } t \in [0,1], \ \ \alpha > 0.$$

If $\alpha = 0$, and $t \in [0,1]$, we obtain:

$$\left| Q_n(t) \right| \leq \| Q_{n-1} \| \left| 1 - \frac{q}{\alpha_n} \right| (1 - \alpha^n) \leq \| Q_{n-1} \| | \left| 1 - \frac{q}{\alpha_n} \right|,$$

$$\| Q_n(t) \|_{C[0,1]} \leq \left| 1 - \frac{q}{\alpha_n} \right| \| Q_{n-1} \|_{C[0,1]}, \ \ n = 2, 3, \ldots,$$

or

$$|\alpha_n - q| t^{\alpha_n} \int_t^1 u^{-1-\alpha_n} du = \frac{|\alpha_n - q|}{\alpha_n} (1 - t^{\alpha_n}) \leq \left| 1 - \frac{q}{\alpha_n} \right|,$$

$$\| Q_0(t) \| = 1,$$

$$\| Q_n(t) \| \leq \prod_{k=1}^{n} \left| 1 - \frac{q}{\alpha_k} \right|,$$

$$\ln \| Q_n(t) \| = \sum_{k=1}^{n} \ln \left| 1 - \frac{q}{\alpha_k} \right|.$$

But, for $\alpha_k \to \infty$, we obtain:

$$\ln \left| 1 - \frac{q}{\alpha_k} \right| \approx -\frac{q}{\alpha_k},$$

$$\ln \| Q_n(t) \| \to -\infty \ \ \Rightarrow \ \ \| Q_n(t) \| \to 0.$$

Thus for all $q \in \mathbb{N}$, $Er(t^q, p(t))_{C[0,1])}$ converges to zero as $n \to \infty$.

Remark We consider the functional

$$I = \frac{1}{\Gamma(\alpha+1)} \int_0^1 \left[t^q - \sum_{k=0}^{n} C_k t^{\alpha_k} \right]^2 (dt)^\alpha \to \min,$$

and noting that

$$\int_0^1 t^{\alpha_i} t^{\alpha_j} (dt)^\alpha = \frac{\Gamma(\alpha_i + \alpha_j + 1)}{\Gamma(\alpha_i + \alpha_j + \alpha + 1)}$$

we have the matrix (see Theorem 7):

$$H(\alpha_1, \alpha_2, \ldots, \alpha_n) = \begin{pmatrix} \frac{\Gamma(\alpha_1+\alpha_1+1)}{\Gamma(\alpha_1+\alpha_1+1+\alpha)} & \cdots & \frac{\Gamma(\alpha_1+\alpha_n+1)}{\Gamma(\alpha_1+\alpha_n+1+\alpha)} \\ & \cdots & \\ \frac{\Gamma(\alpha_n+\alpha_1+1)}{\Gamma(\alpha_n+\alpha_1+1+\alpha)} & \cdots & \frac{\Gamma(\alpha_n+\alpha_n+1)}{\Gamma(\alpha_n+\alpha_n+1+\alpha)} \end{pmatrix}.$$

We cannot calculate $C_1, C_2, \ldots, C_n$ and fractional error $\mathrm{Er}(t^q, p(t))$, in $L^2_\alpha[0, 1]$).

Example To approximate the function

$$f(t) = t^{1/3}, \quad t \in [0, 1],$$

using the function system:

$$\{t^{1/5}, \ t^{1/2}, \ t^{3/4}\}.$$

Solution We will introduce the matrix $A = [a_{ij}]$, where

$$a_{ij} = \int_0^1 t^a t^b dt, \qquad a, b \in \{1/5, 1/2, 3/4\} \qquad i, j = 1 \ldots 3,$$

in order to write the following Maple application for calculation of the a_{ij} elements.

MAPLE

```
with(Student[LinearAlgebra]):
Digits:=4:
A:=≪ 5/7, 10/17, 20/39 > | < 10/17, 1/2, 4/9 > | < 20/39, 4/9, 2/5 ≫;
b:=≪ 15/23 > | < 6/11 > | < 12/25 ≫;
C:= evalf(LeastSquares(A,b));
f(t):=t^(1/3);
g(t):=0.3335*t^(1/5) + 0.9676*t^(1/2) - 0.3027*t^(3/4);
plot(f(t),g(t),t=0..1);
```

```
G:=abs(f(t) - g(t));
for t from 0 to 1 by 0.2 do G od;
```

It results:

$$C_1 = 0.3335, \quad C_2 = 0.9676, \quad C_3 = -0.3027$$

$$|f(0) - g(0)| = 0, \quad |f(0.2) - g(0.2)| = 0.00094$$

$$|f(0.4) - g(0.4)| = 0.0005, \quad \ldots, |f(1) - g(1)| = 0.001$$

Error of Type α

$$Matrix\ A = (a_{ij})_{1 \le i,j \le 3}$$

$$a_{11} = \int_0^1 t^{2/5}(dt)^\alpha = \frac{\Gamma(2/5+1)}{\Gamma(2/5+1+\alpha)}$$

$$a_{12} = \int_0^1 t^{1/5+1/2}(dt)^\alpha = \frac{\Gamma(1/5+1/2+1)}{\Gamma(1/5+1/2+1+\alpha)}$$

$$a_{13} = \int_0^1 t^{1/5+3/4}(dt)^\alpha = \frac{\Gamma(1/5+3/4+1)}{\Gamma(1/5+3/4+1+\alpha)}$$

$$a_{21} = a_{12}$$

$$a_{22} = \int_0^1 t(dt)^\alpha = \frac{\Gamma(2)}{\Gamma(2+\alpha)}$$

$$a_{23} = \int_0^1 t^{1/2+3/4}(dt)^\alpha = \frac{\Gamma(1/2+3/4+1)}{\Gamma(1/2+3/4+1+\alpha)}$$

$$a_{31} = a_{13}$$

$$a_{32} = a_{23}$$

$$a_{33} = \int_0^1 t^{3/2}(dt)^\alpha = \frac{\Gamma(3/2+1)}{\Gamma(3/2+1+\alpha)}$$

$$b_1 = \int_0^1 t^{1/3+1/5}(dt)^\alpha = \frac{\Gamma(1/3+1/5+1)}{\Gamma(1/3+1/5+1+\alpha)}$$

$$b_2 = \int_0^1 t^{1/3+1/2}(dt)^\alpha = \frac{\Gamma(1/3+1/2+1)}{\Gamma(1/3+1/2+1+\alpha)}$$

$$b_3 = \int_0^1 t^{1/3+3/4}(dt)^\alpha = \frac{\Gamma(1/3+3/4+1)}{\Gamma(1/3+3/4+1+\alpha)}$$

For $b = (b_1, b_2, b_3)$, $C = (C_1, C_2, C_3)$ we have

$$AC^T = b^T \Rightarrow C^T = A^{-1}b^T$$

MAPLE

```
restart;
Digits: = 4;
a11:= evalf(GAMMA(2/5 + 1)/GAMMA(2/5 + 1 + a));
a12:= evalf(GAMMA(1/5 + 1/2 + 1)/GAMMA(1/5 + 1/2 + 1 + a);
a13:= evalf(GAMMA(1/5 + 3/4 + 1)/GAMMA(1/5 + 3/4 + 1 + a);
a21:= a12;
a22:= evalf(A(2 + a))
a23:= evalf(1/2 + 3/4 + 1)/GAMMA(1/2 + 3/4 + 1 + a))
a31:= a13;
a32:= a23;
a33:= evalf(3/2 + 1)/GAMMA(3/2 +1 + a))
b1:= evalf(1/3 + 1/5 + 1)/GAMMA(1/3 + 1/5 + 1 + a))
b2:= evalf(1/3 + 1/2 + 1)/GAMMA(1/3 + 1/2 + 1 + a))
b3:= evalf(1/3 + 3/4 + 1)/GAMMA(1/3 + 3/4 + 1 + a))
ec1:= C1*a11 + C2*a12 + C3*a13 = b1;
ec2:= C1*a21 + C2*a22 + C3*a23 = b2;
ec3:= C1*a31 + C2*a32 + C3*a33 = b3;
solve({ec1,ec2,ec3},{C1,C2,C3});
```

for $\alpha = 2/3$, $C_1 = 0.5444$, $C_2 = 0.3366$, $C_3 = 0.1264$

```
f(t):= t^(1/3);
g(t):= 0.54448*t^(1/5) + 0.3366*t^(1/2) + 0.1264*t^(3/4);
plot(f(t),g(t),t = 0..1);
G:= abs(f(t) - g(t));
for t from 0 to 1 by 0.2 do G od;
```

From the last program, we obtain:

$$|f(0) - g(0)| = 0, \quad |f(0.2) - g(0.2)| = 0.001,$$

$$|f(0.4) - g(0.4)| = 0.007, \quad |f(0.6) - g(0.6)| = 0.005,$$

$$|f(0.8) - g(0.8)| = 0.004 \quad |f(1) - g(1)| = 0.007.$$

Alternatively, in Mathematica we have:

MATHEMATICA

```
a11= Gamma(2/5 + 1)/Gamma(2/5 + 1 + a)//N
a12= Gamma(1/5 + 1/2 + 1)/Gamma(1/5 + 1/2 + 1 + a)//N
```

```
a13= Gamma(1/5 + 3/4 + 1)/Gamma(1/5 + 3/4 + 1 + a)//N
a21= a12;
a22= 1/Gamma(2 + a)//N
a23= Gamma(1/2 + 3/4 + 1)/Gamma(1/2 + 3/4 + 1 + a)//N
a31= a13
a32= a23
a33= Gamma(3/2 + 1)/Gamma(3/2 +1 + a)//N
b1= Gamma(1/3 + 1/5 + 1)/Gamma(1/3 + 1/5 + 1 + a)//N
b2= Gamma(1/3 + 1/2 + 1)/Gamma(1/3 + 1/2 + 1 + a)//N
b3= Gamma + 3/4 + 1)/Gamma + 3/4 + 1 + a)//N
ec1= C1*a11 + C2*a12 + C3*a13
ec2= C1*a21 + C2*a22 + C3*a23
ec3= C1*a31 + C2*a32 + C3*a33
Solve[{ec1,ec2,ec3}== {b1,b2,b3},{C1,C2,C3}]//N
f[t]= t^(1/3)
g[t]= 0.54448*t^(1/5) + 0.3366*t^(1/2) + 0.1264*t^(3/4)
Plot[{f[t],g[t]},{t,0,1}]
```

Theorem 8 *If* $\sum_{k=0}^{\infty} C_k t^{k\alpha}$ *is convergent for* $t = t_0$*, then it is convergent whenever* $0 \le t < t_0$.

Proof Suppose that $\sum_{k=0}^{\infty} C_k t_0^{k\alpha}$ is convergent.

Then sequence $\{a_k t_0^{k\alpha}\} \to 0$, for $k \to \infty$.
Thus, there is a constant $M > 0$, so that:

$$\left| C_k t_0^{k\alpha} \right| \le M, \quad k = 0, 1, \ldots$$

Then

$$\left| C_k t^k \right| = \left| C_k t_0^k \right| \left| \frac{t}{t_0} \right| \le M \left| \frac{t}{t_0} \right|^{k\alpha}.$$

Again, if $0 \le t < t_0$, then

$$\left| \frac{t}{t_0} \right|^{k\alpha} < 1,$$

so $\sum_{k=0}^{\infty} \left| \frac{t}{t_0} \right|^{k\alpha}$ is a convergent series.

Applying the comparison test, the series $\sum_{k=0}^{\infty} |C_k t^{k\alpha}|$ is convergent.

Then $\sum_{k=0}^{\infty} C_k t^{k\alpha}$ is absolutely convergent and therefore convergent.

Remark If $\sum_{k=0}^{\infty} C_k t^{k\alpha}$, diverges for $t = t_0$, then it diverges for $t > t_0$.

Example 1 Let us determine the solution of the FDE:

$$D^{(2\alpha)} y(t) + \omega^2 y = 0, \quad 0 < \alpha \leq 1, \quad t > 0,$$

with the conditions:

$$y(0) = A, \qquad D^{(\alpha)} y(0) = B,$$

where A, B are constants. We apply now the series of powers method.

Solution We can write the solution as the series:

$$y(t) = \sum_{n=0}^{\infty} C_n t^{n\alpha}.$$

We have:

$$D^{(2\alpha)} y(t) = \sum_{n=0}^{\infty} C_n D^{(2\alpha)}(t^{n\alpha}) = \sum_{n=2}^{\infty} C_n \frac{\Gamma(n\alpha + 1)}{\Gamma((n-2)\alpha + 1)} t^{(n-2)\alpha}$$

$$= \sum_{n=0}^{\infty} C_{n+2} \frac{\Gamma((n+2)\alpha + 1)}{\Gamma(n\alpha + 1)} t^{n\alpha}.$$

Replacing in the equation, we have:

$$\sum_{n=0}^{\infty} C_{n+2} \frac{\Gamma((n+2)\alpha + 1)}{\Gamma(n\alpha + 1)} t^{2n\alpha} + \omega^2 \sum_{n+0}^{\infty} C_n t^{n\alpha} = 0.$$

Hence:

$$C_{n+2} = -\frac{\omega^2 \Gamma(n\alpha + 1)}{\Gamma((n+2)\alpha + 1)} C_n, \quad n = 0, 1, \ldots,$$

$$C_0 = A, \quad C_1 = \frac{B}{\Gamma(\alpha + 1)}, \quad C_2 = -\frac{\omega^2 A}{\Gamma(2\alpha + 1)}, \quad \ldots$$

$$y(t) = A \sum_{n=0}^{\infty} \frac{(-1)^n \omega^{2n}}{\Gamma(2n\alpha + 1)} t^{2n\alpha} + B \sum_{n=0}^{\infty} \frac{(-1)^n \omega^{2n}}{\Gamma((2n+1)\alpha + 1)} t^{(2n+1)\alpha},$$

or:

$$y(t) = AE_{2\alpha}(-\omega^2 t^{2\alpha}) + B\sum_{n=0}^{\infty}\frac{(-1)^n\omega^{2n}}{\Gamma((2n+1)\alpha+1)}t^{(2n+1)\alpha}.$$

Example 2 Let us determine the solution of the FDE [2]:

$$D^{(\alpha)}y(t) - y^2 - 1 = 0, \;\; m-1 < \alpha \le m, \;\; t > 0,$$

with the initial conditions

$$y^{(i)}(0) = 0, \qquad i = 0, 1, \ldots, m-1,$$

using the series of powers method.

Solution We take the solution as a series of powers

$$y(t) = \sum_{n=0}^{\infty} C_n t^{n\alpha}.$$

We define the so-called αk-th FD as

$$D^{(k\alpha)}(D^{(\alpha)}y(t) - y^2(t) - 1) = 0, \;\; k = 0, 1, \ldots, m-1,$$

and introducing the solution in the equation we obtain

$$D^{(\alpha(k+1))}\left(\sum_{n=0}^{\infty} C_n t^{n\alpha}\right) - D^{(\alpha k)}\left(\sum_{n=0}^{\infty} C_n t^{n\alpha}\right) = D^{(\alpha k)}(1), \qquad k = 0, 1, \ldots,$$

$$\sum_{n=k+1}^{\infty} C_n \frac{\Gamma(n\alpha+1)}{\Gamma((n-k-1)\alpha+1)}t^{(n-k-1)\alpha} - \sum_{n=k}^{\infty}\left(\sum_{j=0}^{n} C_j C_{n-j}\right) \times \frac{\Gamma(n\alpha+1)}{\Gamma((n-k-1)\alpha+1)}t^{(n-k)\alpha} = \chi_k,$$

where

$$\chi_k = \begin{cases} 1, & k = 0, \\ 0, & k \ge 1. \end{cases}$$

By identification, we get the coefficients

$$C_0 = 0,$$

$$C_1 = \frac{1}{\Gamma(\alpha+1)},$$

$$C_{k+1} = \frac{\Gamma(k\alpha+1)}{\Gamma((1+k)\alpha+1)} \sum_{j=0}^{n} C_j C_{k-j},$$

and the solution:

$$y(t) = \frac{1}{\Gamma(\alpha+1)} t^{\alpha} + \frac{\Gamma(2\alpha+1)}{\Gamma^2(\alpha+1)\Gamma(3\alpha+1)} t^{3\alpha} + \ldots$$

5.2.2 Lane-Emden Equation

We apply now the series of powers method to solve the FDE of Lane[4] and Emden[5] [5]

$$D^{\alpha} y(t) + \frac{a_1}{t^{\alpha-\beta_1}} D^{\beta_1} y(t) + \frac{a_2}{t^{\alpha-\beta_2}} D^{\beta_2} y(t) + \ldots + \frac{a_n}{t^{\alpha-\beta_n}} D^{\beta_n} y(t) + y^m(t) = 0,$$

and the initial conditions are:

$$y(0) = 1, \qquad y'(0) = 0.$$

We consider that $0 < t \le 1$, $0 < \beta_i \le 1$, $i = 1, 2, \ldots, n$, $1 < \alpha \le 2$, $a_i \in \mathbb{R}$, and $m \in \mathbb{N}_+$. The equation can be written also as $Ly + y^m = 0$, using the linear operator L.

Solution We take the approximate solution as a series of powers:

$$y(t) = \sum_{k=0}^{\infty} C_k t^{k\alpha},$$

where C_k are constants. The term Ly can be written also as:

$$L[y(t)] = D^{\alpha} y + \sum_{i=0}^{n} \frac{a_i}{t^{\alpha-\beta_i}} D^{\beta_i} y(t).$$

[4]J.H. Lane (1819–1880).

[5]J.R. Emden (1862–1940).

We will use also the formula:

$$D^{\alpha}t^{\lambda} = \frac{\Gamma(\lambda+1)}{\Gamma(\lambda+1-\alpha)}t^{\lambda-\alpha}.$$

It results:

$$D^{\alpha}y = \sum_{k=1}^{\infty} C_k \frac{\Gamma(\alpha k+1)}{\Gamma(\alpha k+1-\alpha)}t^{k\alpha-\alpha} = \sum_{k=0}^{\infty} C_{k+1}\frac{\Gamma(\alpha k+1+\alpha)}{\Gamma(\alpha k+1)}t^{k\alpha}$$

$$\begin{aligned}\frac{a_i}{t^{\alpha-\beta_i}}D^{\beta_i}y(t) &= \frac{a_i}{t^{\alpha-\beta_i}}\sum_{k=1}^{\infty} C_k \frac{\Gamma(k\alpha+1)}{\Gamma(k\alpha+1-\beta_i)}t^{k\alpha-\beta_i}\\ &= \sum_{k=0}^{\infty} C_{k+1}\, a_i \frac{\Gamma(k\alpha+1+\alpha)}{\Gamma(k\alpha+1+\alpha-\beta_i)}t^{k\alpha}.\end{aligned}$$

The equation becomes:

$$\sum_{k=0}^{\infty} C_{k+1}\Gamma(\alpha k+1+\alpha)\left[\frac{1}{\Gamma(\alpha k+1)} + \sum_{i=1}^{n}\frac{a_i}{\Gamma(k\alpha+1+\alpha-\beta_i)}\right]t^{\alpha k} + y^m = 0.$$

We introduce now the notation $F(\alpha, k)$:

$$F(\alpha, k) = C_{k+1}\Gamma(\alpha k+1+\alpha)\left[\frac{1}{\Gamma(\alpha k+1)} + \sum_{i=1}^{n}\frac{a_i}{\Gamma(k\alpha+1+\alpha-\beta_i)}\right].$$

We will examine some cases:

(a) In the case $m = 0$, we have:

$$\sum_{k=0}^{\infty} F(\alpha k)t^{\alpha k} + 1 = 0.$$

Then:

$$C_0 = 1,$$

$$C_1\Gamma(\alpha+1)\left[1 + \sum_{i=1}^{n}\frac{a_i}{\Gamma(1+\alpha-\beta_i)}\right] + 1 = 0,$$

$$C_i = 0, \qquad i = 2, 3, \ldots$$

For $\alpha = 2, \quad \beta = 1, \quad a_1 = 2, a_i = 0, \quad i = 2, 3, \ldots$

$$y''(t) + \frac{2}{t}y' + 1 = 0,$$

resulting the solution:

$$y(t) = 1 - \frac{t^2}{6}.$$

(b) The case $m > 0$.

Then we have the following relations for different values of m:

1. $m = 1, \qquad F(\alpha, k) + C_k = 0,$
2. $m = 2, \qquad F(\alpha, k) + \sum_{i=0}^{k} C_i C_{k-i} = 0$, because:

$$y^2(t) = \left(\sum_{k=0}^{\infty} C_k t^{\alpha k}\right)^2 = \sum_{k=0}^{\infty}\sum_{i=0}^{k} C_i C_{k-i} t^{\alpha k}.$$

3. $m = 3, \qquad F(\alpha, k) + \sum_{i=0}^{k}\sum_{j=0}^{k-i} C_i C_j C_{k-i-j} = 0$, because:

$$y^3(t) = \left(\sum_{k=0}^{\infty} C_k t^{\alpha k}\right)^3 = \sum_{k=0}^{\infty}\sum_{i=0}^{k}\sum_{j=0}^{k-i} C_i C_j C_{k-i-j} t^{\alpha k},$$

or, the relation:

$$F(\alpha, k) + \sum_{i=0}^{k}\sum_{j=0}^{k-i} \cdots \sum_{t_m=0}^{k-i-j-\ldots t_{m-1}} C_i C_j \ldots C_{k-i-\ldots-t_m} = 0.$$

For same special cases, we have:

$$\alpha = 2, \quad \beta = 1, \quad a_1 = 2, \quad a_i = 0, \quad i = 2, 3 \ldots,$$

$(a)\ m = 1,\ y''(t) + \frac{2}{t}y'(t) + y(t) = 0 \Leftrightarrow y(t) = 1 - \frac{t^2}{6} + \ldots,$

$(b)\ m = 2,\ y''(t) + \frac{2}{t}y'(t) + y^2(t) = 0 \Leftrightarrow y(t) = 1 - \frac{t^2}{3!} + \frac{t^4}{5!} + \ldots = \frac{\sin t}{t},$

$$(c)\ \ m = 3,\ y''(t) + \frac{2}{t}y'(t) + y^3(t) = 0 \Leftrightarrow y(t) = 1 - \frac{t^2}{3!} + \frac{t^4}{60} + \ldots$$

These results are confirmed by the following Maple program:

MAPLE

```
>m:= ... is introduced here
>ec:= diff(y(t),t,t) + 2/t*diff(y(t),t)+(y(t))^m = 0;
>dsolve({ec, y(0) = 1, D(y)(0) = 0}, y(t), series);
```

Other results are as follows:

$$(1)\, m = 4, \qquad y(t) = 1 - \frac{t^2}{6} + \frac{t^4}{30} + O(t^6),$$

$$(2)\, m = 5, \qquad y(t) = 1 - \frac{t^2}{6} + \frac{t^4}{24} + O(t^6),$$

$$(3)\, m = 6, \qquad y(t) = 1 - \frac{t^2}{6} + \frac{t^4}{20} + O(t^6),$$

$$(4)\, m = 10, \qquad y(t) = 1 - \frac{t^2}{6} + \frac{t^4}{12} + O(t^6),$$

$$(5)\, m = 20, \qquad y(t) = 1 - \frac{t^2}{6} + \frac{t^4}{6} + O(t^6).$$

Finally we will examine the following particular case:

$$y^{(\frac{3}{2})}(t) + \frac{2}{t}y^{(\frac{1}{2})}(t) + \frac{1}{t^{\frac{3}{4}}}y^{(\frac{3}{4})}(t) + y^m(t) = 0,$$

$$\alpha = \frac{3}{2},\ \ \beta_1 = \frac{1}{2},\ \ \beta_2 = \frac{3}{4},\ \ a_1 = 2,\ \ a_2 = 1,$$

$$G(k) = F\left(\frac{3}{2}, k\right) = \Gamma\left(\frac{3}{2}k + \frac{5}{2}\right)\left[\frac{1}{\Gamma\left(\frac{3}{2}k + 1\right)} + \frac{2}{\Gamma\left(\frac{3}{2}k + 2\right)} + \frac{1}{\Gamma\left(\frac{3}{2}k + \frac{3}{4}\right)}\right],$$

$$G(0) = 5.07282,\ \ G(1) = 13.4199,\ \ G(2) = 24.9119, \ldots$$

$$(a)\, m=0, \qquad \sum_{k=0}^{\infty} G(k)t^{\frac{3}{2}k}+1=0$$

$$C_0=1,\ \ C_1=-\frac{1}{G(0)}=-0.19742,\ \ C_i=0,\ \ i=2,3,\ldots,$$

$$y(t)=1-0.1971\,t^{\frac{3}{2}},$$

$$(b)\ \ m=1, \qquad C_{k+1}G(k)+C_k=0,$$

$$C_0=1,\ \ ,C_1=-\frac{1}{G(0)}=-0.19742,$$

$$C_2G(1)+C_1=0\ \Rightarrow\ C_2=0.0146,\ldots,$$

$$y(t)=1-0.1971t^{\frac{3}{2}}+0.0146t^3+\ldots,$$

$$(c)\ \ m=2, \qquad C_{k+1}G(k)+\sum_{i=0}^{k} C_iC_{k-i}=0,$$

$$C_0=1,$$

$$C_1G(0)+C_0^2=0\ \Rightarrow\ C_1=-0.1971,$$

$$C_2G(1)+2C_0C_1=0\ \Rightarrow\ C_2=0.0293,$$

$$y(t)=1-0.1971t^{\frac{3}{2}}+0.0293t^3+\ldots$$

5.2.3 The Taylor Series Method

Theorem 9 *The series* $\sum_{k=0}^{\infty} C_k t^{k\alpha}$, *here* $C_0 \neq 0$, *is convergent for* $0<t<R^{1/\alpha}$.

Proof We can write:

$$\left|\frac{C_{n+1}t^{(n+1)\alpha}}{C_n t^{n\alpha}}\right|<1,$$

or:

$$0<t^{\alpha}<\left|\frac{C_n}{C_{n+1}}\right|.$$

But the convergence radius is $R = \lim_{n\to\infty} \left| \frac{C_n}{C_{n+1}} \right|$ so that, finally, we obtain:

$$0 < t < R^{1/\alpha}.$$

Theorem 10 *Suppose that:*

$$f(t) = \sum_{n=0}^{\infty} C_n t^{n\alpha}, \qquad 0 \le m-1 < \alpha \le m, \qquad 0 \le t < R^{1/\alpha}.$$

If $f(t) \in C[0, R^{1/\alpha})$, *and* $D^{(n\alpha)} f(t) \in C(0, R^{1/\alpha})$, *for* $n = 0, 1, \ldots$, *then:*

$$C_n = \frac{D^{n\alpha} f(0)}{\Gamma(n\alpha+1)},$$

where $D^{(n\alpha)} = D^{(\alpha)} D^{(\alpha)} \ldots D^{(\alpha)} = (D^{(\alpha)})^n$.

Proof For $t = 0$, we have $C_0 = f(0)$.

Using the formula

$$D^{(n\alpha)} t^{\lambda} = \frac{\Gamma(\lambda+1)}{\Gamma(\lambda+1-n\alpha)} t^{\lambda-n\alpha},$$

it results:

$$D^{(\alpha)} f(t) = C_1 \Gamma(\alpha+1) + C_2 \frac{\Gamma(2\alpha+1)}{\Gamma(\alpha+1)} t^{\alpha} + C_3 \frac{\Gamma(3\alpha+1)}{\Gamma(2\alpha+1)} t^{2\alpha} + \ldots$$

For:

$$t = 0, \qquad \text{we have: } C_1 = \frac{D^{(\alpha)} f(0)}{\Gamma(\alpha+1)}.$$

However, if we continue to apply n times $D^{(\alpha)}$ we obtain for $t = 0$:

$$C_n = \frac{D^{(n\alpha)} f(0)}{\Gamma(n\alpha+1)}.$$

Thus we have the generalized MacLaurin series:

$$f(t) = \sum_{n=0}^{\infty} \frac{D^{(n\alpha)} f(0)}{\Gamma(n\alpha+1)} t^{n\alpha}.$$

Remark Similarly, for $t_0 \le t < R^{1/\alpha}$ and

$$f(t) = \sum_{n=0}^{\infty} (t - t_0)^{n\alpha},$$

we have the Taylor generalized formula:

$$f(t) = \sum_{n=0}^{\infty} \frac{D_{t_0}^{(n\alpha)} f(t_0)}{\Gamma(n\alpha + 1)} (t - t_0)^{n\alpha}.$$

5.2.4 *The Generalized Hermite Equation*

The Hermite[6] equation [6] can be generalized as [4]:

$$D^{(2\alpha)} y(t) - 2t^{\alpha} D^{(\alpha)} y(t) + \lambda y(t) = 0,$$

for the fractional case.

Solution The solution can be written using the powers series method, as:

$$y(t) = \sum_{n=0}^{\infty} C_n t^{n\alpha}, \qquad (C_0 \neq 0),$$

where C_n are constants.

In the calculations we apply the formula:

$$D^{\alpha} t^{\lambda} = \frac{\Gamma(\lambda + 1)}{\Gamma(\lambda + 1 - \alpha)} t^{\lambda - \alpha}.$$

It results:

$$\sum_{n=2}^{\infty} C_n D^{(2\alpha)} t^{n\alpha} - 2t^{\alpha} \sum_{n=1}^{\infty} C_n D^{(\alpha)} t^{n\alpha} + \lambda \sum_{n=0}^{\infty} C_n t^{n\alpha} = 0,$$

$$\sum_{n=2}^{\infty} C_n \frac{\Gamma(n\alpha + 1)}{\Gamma((n-2)\alpha + 1)} t^{(n-2)\alpha} - 2t^{\alpha} \sum_{n=1}^{\infty} C_n \frac{\Gamma(n\alpha + 1)}{\Gamma((n-1)\alpha + 1)} t^{(n-1)\alpha}$$
$$+ \lambda \sum_{n=0}^{\infty} C_n t^{n\alpha} = 0,$$

[6]C. Hermite (1822–1901).

or, finally:

$$\sum_{n=0}^{\infty} C_{n+2}\frac{\Gamma((n+2)\alpha+1)}{\Gamma(n\alpha+1)}t^{n\alpha} - 2\sum_{n=1}^{\infty} C_n\frac{\Gamma(n\alpha+1)}{\Gamma((n-1)\alpha+1)}t^{n\alpha} + \lambda\sum_{n=0}^{\infty} C_n t^{n\alpha} = 0$$

For $n = 0 \Rightarrow C_1\Gamma(2\alpha+1) + \lambda C_0 = 0 \quad \Rightarrow \quad C_1 = -\dfrac{\lambda C_0}{\Gamma(2\alpha+1)}$

$$C_{m+1}\frac{\Gamma((n+2)\alpha+1)}{\Gamma(n\alpha+1)} = C_n\left[\frac{2\Gamma(n\alpha+1)}{\Gamma((n-1)\alpha+1)} - \lambda\right].$$

For $\alpha = 1$, we obtain:

$$C_1 = -\frac{\lambda C_0}{2!},$$

$$C_{n+1} = C_n\frac{2n-\lambda}{(n+1)(n+2)},$$

$$y(t) = C_0\left[1 - \frac{1}{2!}\lambda t^2 + \frac{1}{4!}\lambda(\lambda-4)t^3 + \ldots\right].$$

For $\alpha = \dfrac{1}{2}$, we have

$$C_1 = -\lambda C_0, \quad C_2 = \frac{2\lambda C_9}{3}, \quad C_3 = \frac{\lambda^2 C_0}{3}\left(\frac{2}{\sqrt{\pi}} - \lambda\right), \quad \ldots,$$

and the solution:

$$y(t) = C_0\left[1 - \lambda t^{1/2} + \frac{2\lambda}{3}t + \frac{\lambda^2}{3}\left(\frac{2}{\sqrt{\pi}} - \lambda\right)t^{3/2}\right]$$

5.2.5 *The Generalized Legendre Equation*

The generalized Legendre FDE can be defined as [4]:

$$(1 - t^{2\alpha})D^{(2\alpha)}y(t) - 2t^{\alpha}D^{(\alpha)}y(t) + \lambda y(t) = 0.$$

Solution We consider a power series solution:

$$y(t) = \sum_{n=0}^{\infty} C_n t^{n\alpha}, \qquad C_0 \neq 0$$

where C_n are constants. We replace the solution in the equation

$$(1-t^{2\alpha})\sum_{n=2}^{\infty} C_n D^{(2\alpha)} t^{\alpha n} - 2t^{\alpha}\sum_{n=1}^{\infty} C_n D^{(\alpha)} t^{\alpha n} + \lambda \sum_{n=0}^{\infty} C_n t^{\alpha n} = 0,$$

or:

$$\sum_{n=2}^{\infty} C_n \frac{\Gamma(n\alpha+1)}{\Gamma((n-2)\alpha+1)} t^{(n-2)\alpha} - \sum_{n=2}^{\infty} C_n \frac{\Gamma(n\alpha+1)}{\Gamma((n-2)\alpha+1)} t^{n\alpha}$$
$$-2\sum_{n=1}^{\infty} C_n \frac{\Gamma(n\alpha+1)}{\Gamma((n-1)\alpha+1)} t^{n\alpha}$$
$$+\lambda \sum_{n=0}^{\infty} C_n t^{n\alpha} = 0,$$

and, finally we obtain:

$$ds \sum_{n=0}^{\infty} C_{n+2} \frac{\Gamma((n+2)\alpha+1)}{\Gamma(n\alpha+1)} t^{n\alpha} - \sum_{n=2}^{\infty} C_n \frac{\Gamma(n\alpha+1)}{\Gamma((n-2)\alpha+1)} t^{n\alpha}$$
$$-2\sum_{n=1}^{\infty} C_n \frac{\Gamma(n\alpha+1)}{\Gamma((n-1)\alpha+1)} t^{n\alpha}$$
$$+\lambda \sum_{n=0}^{\infty} C_n t^{n\alpha} = 0.$$

For $n = 0$, we have:

$$C_2\Gamma(2\alpha+1) + \lambda C_0 = 0, \quad \Rightarrow \quad C_2 = -\frac{C_0\lambda}{\Gamma(2\alpha+1)}.$$

For $n = 1$, we obtain:

$$C_3 \frac{\Gamma(3\alpha+1)}{\Gamma(\alpha+1)} - 2C_1\Gamma(\alpha+1) + \lambda C_1 = 0, \quad \Rightarrow \quad C_3 = -C_1 \frac{\lambda - 2\Gamma(\alpha+1)}{\Gamma(3\alpha+1)} \Gamma(\alpha+1)$$

$$C_{n+2} \frac{\Gamma(\alpha(n+2)+1)}{\Gamma(\alpha n+1)} - C_n \frac{\Gamma(\alpha n+1)}{\Gamma(\alpha(n-2)+1)}$$
$$-2C_n \frac{\Gamma(\alpha n+1)}{\Gamma(\alpha(n-1)+1)} + \lambda C_n = 0.$$

For $\alpha = 1$

$$C_2 = -\frac{\lambda C_0}{6},$$

$$C_3 = -C_1\frac{\lambda - 2}{6},$$

$$C_4 = C_0\frac{\lambda(\lambda - 6)}{24},$$

$$\dots$$

and the solution:

$$y(t) = C_0\left[1 - \frac{\lambda}{2}t^2 + \frac{\lambda(\lambda - 6)}{24}t^4 + \dots\right] + C_1\left[t - \frac{\lambda - 2}{6}t^3 + \dots\right]$$

For $\alpha = \frac{1}{2}$, we have:

$$C_2 = -\frac{\lambda C_0}{6},$$

$$C_3 = -2C_1\frac{\lambda - \sqrt{\pi}}{3},$$

$$C_4 = -C_0\frac{1}{12}\left[1 - \lambda + \frac{4}{\sqrt{\pi}}\right].$$

5.2.6 *The Generalized Bessel Equation*

The Bessel[7] FDE can be introduced as [4]:

$$t^{2\alpha}D^{(2\alpha)}y(t) + t^{\alpha}D^{(\alpha)}y(t) + (t^{2\alpha} - p^2)y(t) = 0, \quad p \in R.$$

Solution We consider a power series solution:

$$y(t) = \sum_{n=0}^{\infty} C_n t^{\lambda + n\alpha}, \qquad (C_0 \neq 0),$$

[7]F. Bessel (1784–1846).

where C_n are constants. Replacing the solution in the equation, we have:

$$t^{2\alpha} D^{(2\alpha)} y(t) = t^{2\alpha} \sum_{n=0}^{\infty} C_n \frac{\Gamma(\lambda + n\alpha + 1)}{\Gamma(\lambda + (n-2)\alpha + 1)} t^{\lambda + n\alpha - 2\alpha}$$
$$= \sum_{n=0}^{\infty} C_n \frac{\Gamma(\lambda + n\alpha + 1)}{\Gamma(\lambda + (n-2)\alpha + 1)} t^{\lambda + n\alpha},$$

or

$$t^{\alpha} D^{(\alpha)} y(t) = t^{\alpha} \sum_{n=0}^{\infty} C_n \frac{\Gamma(\lambda + n\alpha + 1)}{\Gamma(\lambda + (n-1)\alpha + 1)} t^{\lambda + n\alpha - \alpha}$$
$$= \sum_{n=0}^{\infty} C_n \frac{\Gamma(\lambda + n\alpha + 1)}{\Gamma(\lambda + (n-1)\alpha + 1)} t^{\lambda + n\alpha},$$

and, finally:

$$(t^{2\alpha} - p^2) \sum_{n=0}^{\infty} C_n t^{\lambda + n\alpha} = \sum_{n=0}^{\infty} C_n t^{\lambda + (n+2)\alpha} - p^2 \sum_{n=0}^{\infty} C_n t^{\lambda + n\alpha}$$
$$= \sum_{n=0}^{\infty} C_{n-2} t^{\lambda + n\alpha} - p^2 \sum_{n=0}^{\infty} C_n t^{\lambda + n\alpha}.$$

Hence, the recurrence relation between the coefficients C_n is:

$$\sum_{n=0}^{\infty} C_n \frac{\Gamma(\lambda + n\alpha + 1)}{\Gamma((n-2)\alpha + 1)} t^{\lambda + n\alpha} + \sum_{n=0}^{\infty} C_n \frac{\Gamma(\lambda + n\alpha + 1)}{\Gamma((n-1)\alpha + 1)} t^{\lambda + n\alpha}$$
$$+ \sum_{n=0}^{\infty} C_{n-2} t^{\lambda + n\alpha} - p^2 \sum_{n=0}^{\infty} C_n t^{\lambda + n\alpha} = 0.$$

From the recurrence relation, it results:

$$C_0 \left[\frac{\Gamma(\lambda + 1)}{\Gamma(\lambda - 2\alpha + 1)} + \frac{\Gamma(\lambda + 1)}{\Gamma(\lambda - \alpha + 1)} - p^2 \right] = 0,$$

$$C_1 \left[\frac{\Gamma(\lambda + \alpha + 1)}{\Gamma(\lambda - \alpha + 1)} + \frac{\Gamma(\lambda + \alpha + 1}{\Gamma(\lambda + 1)} - p^2 \right] = 0,$$

$$\ldots,$$

$$C_k \left[\frac{\Gamma(\lambda + k\alpha + 1}{\Gamma(\lambda + (k-2)\alpha + 1)} + \frac{\Gamma(\lambda + k\alpha + 1)}{\Gamma(\lambda + 1)} - p^2 \right] + C_{k-2} = 0.$$

For $\alpha = 1$, we have:

$$C_0\left[\lambda^2 - p^2\right] = 0,$$

$$C_1\left[(\lambda + 1)^2 - p^2\right] = 0.$$

For C_0 arbitrary, $\lambda = \pm p$, $(\lambda + 1 \neq 0)$ results $C_1 = 0$.

$$C_{2k} = -\frac{C_{2k-2}}{(2p + 2k)(2k)},$$

or:

$$C_{2k} = (-1)^{k+1}\frac{C_0}{2 \cdot 4 \cdot 6 \cdots 2k\,(2p + 2) \cdots (2p + 2k)}.$$

We obtain the solution:

$$y(t) = C_0 \sum_{k=0}^{\infty} \frac{(-1)^k t^{p+2k}}{4^k k!(p + 1)(p + 2) \cdots (p + k)},$$

and

$$J_p(t) = \frac{1}{2^p \Gamma(p + 1)} \sum_{k=0}^{\infty} \frac{(-1)^k t^{p+2k}}{4^k k!(p + 1)(p + 2) \cdots (p + k)},$$

$$J_p(t) = \sum_{k=0}^{\infty} \frac{(-1)^k}{k!\Gamma(p + k + 1)} \left(\frac{t}{2}\right)^{p+2k},$$

where Bessel function of first kind $(J_p(t))$ is called the Bessel function of first kind.

For $\alpha = \frac{1}{2}$, C_0 arbitrary, $C_1 = 0$, using Mathematica we will solve the equation:

$$\lambda + \frac{\lambda\Gamma(\lambda)}{\Gamma\left(\lambda + \frac{1}{2}\right)} - p^2 = 0.$$

For $p = 2$, if we apply the Mathematica command:

$$FindRoot[x + x * \Gamma[x + 1/2] - 4 == 0, \{x, 0.1\}],$$

we obtain $x = \lambda = 2.37593$, [3].

5.2.7 Nonlinear Systems

Lotka System

We consider as a first example the Lotka[8] system with initial conditions:

$$\begin{cases} D^{\alpha} x(t) = 3.5\, y(t)\, (1 - y(t)), & x(0) = 0.2, \\ D^{\alpha} y(t) = 4\, x(t)\, (1 - x(t)), & y(0) = 0.2. \end{cases}$$

We look here for solutions in the form of series of powers:

$$x(t) = \sum_{n=0}^{\infty} a_n t^{n\alpha}, \qquad y(t) = \sum_{n=0}^{\infty} b_n t^{n\alpha}.$$

We use the derivation rule:

$$D^{\alpha} t^{\lambda} = \frac{\Gamma(\lambda + 1)}{\Gamma(\lambda + 1 - \alpha)} t^{\lambda - \alpha}.$$

It results:

$$D^{\alpha} \left(\sum_{n=0}^{\infty} a_n t^{n\alpha} \right) = \sum_{n=1}^{\infty} a_n D^{\alpha} t^{n\alpha} = \sum_{n=1}^{\infty} \frac{\Gamma(n\alpha + 1)}{\Gamma((n-1)\alpha + 1)}$$

$$= \sum_{m=0}^{\infty} \frac{\Gamma((m+1)\alpha + 1)}{\Gamma(m\alpha + 1)} t^{m\alpha} = \sum_{n=0}^{\infty} \frac{\Gamma((n+1)\alpha + 1)}{\Gamma(n\alpha + 1)} t^{n\alpha}.$$

We obtain the following recurrence relations:

$$\begin{cases} \dfrac{\Gamma((n+1)\alpha + 1)}{\Gamma(n\alpha + 1)} a_{n+1} = 3.5 (b_n - \sum_{k}^{n-k} b_k b_{n-k}), \\ a_0 = 0.2, \\ \dfrac{\Gamma((n+1)\alpha + 1)}{\Gamma(n\alpha + 1)} b_{n+1} = 4 \left(a_n - \sum_{k}^{n-k} a_k a_{n-k} \right), \\ b_0 = 0.2. \end{cases}$$

---

[8] A.J. Lotka (1880–1949).

For numerical calculations of a_n, b_n, we can use the Mathematica program:

MATHEMATICA

```
Clear["`*"]
\[Alpha] := 3/4;
a[0] = 0.1; b[0] = 0;
f[n_] := Gamma[n*\[Alpha] + 1]/Gamma[(n + 1)*
    \[Alpha]+1];
  For[n = 0, n <= 5,
 n++, {a[n + 1] = 3.5*f[n]*(b[n] -
        Sum[b[k]*b[n - k], {k, 0, n}]),
  b[n + 1] = 4*f[n]*(a[n] -
   Sum[a[k]*a[n - k], {k, 0, n}]), {n, 0, 5}}]
TableForm[Table[{n, a[n], b[n]}, {n, 0, 5}]]

0  0.1      0
1  0        0.3917
2  0.9475   1.5816
3 -0.2800  1.58164
4  2.3520  -0.3807
5 -2.0508  1.42210

sol = Table[{a[n], b[n]}, {n, 1, 1000}];
p = Interpolation /@ Transpose@sol;
ParametricPlot[Evaluate@Through@p@t,
        {t, 1, 1000}, Frame -> True]
```

Figure 5.5 shows the solution $(a(t), b(t))$ of the Lotka attractor.

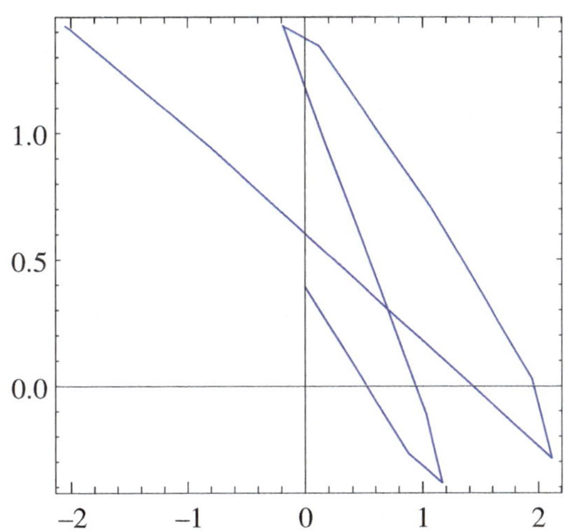

Fig. 5.5 Lotka attractor solution $(a(t), b(t))$

Lorenz Fractional Attractor

The Lorenz[9] fractional attractor is defined by the system of FDE:

$$\begin{cases} D^{\alpha}x(t) = 10\,(y(t) - x(t)), \\ D^{\alpha}y(t) = x(t)\,(28 - z(t)) - y(t), \\ D^{\alpha}z(t) = x(t)\,y(t) - \dfrac{8}{3}z(t). \end{cases}$$

We use the initial conditions:

$$x(0) = 0.1, \quad y(0) = 0.1, \quad z(0) = 0.1.$$

We take the power series solutions and $0 < \alpha \leq 1$:

$$x(t) = \sum_{n=0}^{\infty} a_n t^{n\alpha}, \qquad y(t) = \sum_{n=0}^{\infty} b_n t^{n\alpha}, \qquad z(t) = \sum_{n=0}^{\infty} c_n t^{n\alpha}.$$

We obtain the recurrence relations:

$$\begin{cases} a_{n+1}\dfrac{\Gamma((n+1)\alpha+1)}{\Gamma(n\alpha+1)} = 10[b_n - a_n] \\ b_{n+1}\dfrac{\Gamma((n+1)\alpha+1)}{\Gamma(n\alpha+1)} = 28a_n - \displaystyle\sum_{k=0}^{n} a_k c_{n-k} - b_n \\ c_{n+1}\dfrac{\Gamma((n+1)\alpha+1)}{\Gamma(n\alpha+1)} = \displaystyle\sum_{k=0}^{n} a_k b_{n-k} - \dfrac{8}{3}c_n. \end{cases}$$

Example For $\alpha = 0.95$, we have:

$$\begin{cases} D^{0.95}x(t) = 10[y(t) - x(t)] \\ D^{0.95}y(t) = x(t)[28 - z(t)] - y(t) \\ D^{0.95}z(t) = x(t)y(t) - \dfrac{8}{3}z(t) \end{cases}$$

$$x(0) = 0.1, \qquad y(0) = 0.1, \qquad z(0) = 0.1.$$

[9]E.N. Lorenz (1917–2008).

The coefficients a_n, b_n, c_n can be calculated using the program:

MAPLE

```
restart:Digits:=5:
> a[0]:=0.1:
> b[0]:=0.1:
> c[0]:=0.1:
> alpha:=0.95:
> unassign('n');
> f:=n->GAMMA(n*alpha+1)/GAMMA((n+1)*alpha+1);

                                   GAMMA(n alpha + 1)
                      f := n -> ------------------------
                                GAMMA((n + 1) alpha + 1)

> a[1]:=f(0)*10*(b[0]-a[0]):
> b[1]:=f(0)*(28*a[0]-a[0]*c[0]-b[0]):
> c[1]:=f(0)*(a[0]*b[0]-8/3*c[0]):
> for n from 1 to 5 do
> a[n+1]:=f(n)*10*(b[n]-a[n]);
> b[n+1]:=f(n)*(28*a[n]-sum(a[k]*b[n-k],k=0..n)-b[n]);
> c[n+1]:=f(n)*(sum(a[k]*b[n-k],k=0..n)-8/3*c[n]);
> od;
```

The solutions are:

$$x(t) = 0.1 - 0t^{0.95} + 14.720t^{2 \cdot 0.95} - 59.987t^{3 \cdot 0.95} + 519.94t^{4 \cdot 0.95} -$$

$$-2523.97t^{5 \cdot 0.95} + 12242t^{6 \cdot 0.95} \ldots$$

$$y(t) = 0.1 + 2.7451t^{0.95} - 1.6192t^{2 \cdot 0.95} + 151.18t^{3 \cdot 0.95} - 524.76t^{4 \cdot 0.95} +$$

$$+3900t^{5 \cdot 0.95} - 14988t^{6 \cdot 0.95} \ldots$$

$$z(t) = 0.1 - 0.26193t^{0.95} + 0.52174t^{2 \cdot 0.95} - 0.02977t^{3 \cdot 0.95} +$$

$$+13.868t^{4 \cdot 0.95} - 49.518t^{5 \cdot 0.95} + 802.5t^{6 \cdot 0.95} \ldots$$

For $\alpha = 0.995$, the coefficients a_n, b_n, c_n are:

MATHEMATICA

In this case, we have:

```
Clear["`*"]
\[Alpha] := 0.995;
a[0] = 0.1; b[0] = 0.1; c[0] = 0.1;
```

```
f[n_] := Gamma[n*\[Alpha] + 1]/Gamma[(n + 1)
  *\[Alpha] + 1];
a[1] = f[0]*10*(b[0] - a[0]);
b[1] = f[0]*(28*a[0] - a[0]*c[0] - b[0]);
c[1] = f[0]*(a[0]*b[0] - 8/3*c[0]); For[n = 0, n <= 5,
 n++, {a[n + 1] = 10*f[n]*(b[n] - a[n]),
  b[n + 1] = f[n]*(28*a[n] - Sum[a[k]*b[n - k],
  {k, 0, n}]),
  c[n + 1] = f[n]*(Sum[a[k]*b[n - k], {k, 0, n}]
  - 8/3*c[n])}]
TableForm[Table[{n, a[n], b[n], c[n]}, {n, 0, 5}]]
```

References

1. Almira, J. M. (2007). Müntz type theorem I. *Surveys in Approximation Theory, 3*, 152–194.
2. El-Ajou, A., Arqub, O. A., Al Zhour, Z., & Momani, S. (2013). Application of a variational iteration method to linear and nonlinear viscoelastic models with fractional derivatives. *Entropy, 15*, 5305–5323.
3. Johnson, R. S. (2006). *The notebook series. the series of second order ordinary differential equation and special function*. Technical report, School of Mathematics & Statistics, University of Newcastle upon Tyne.
4. Milici, C., & Drăgănescu, G. (2017). Generalization of the equations of Hermite, Legendre and Bessel for the fractional case. *Journal of Applied Nonlinear Dynamics, 6*, 243–249.
5. Milici, C., & Drăgănescu, G. (2017). The Lane-Emden fractional homogenous differential equation. *Journal of Applied Nonlinear Dynamics, 6*, 237–242.
6. Nikiforov, A. F., & Ouvarov, V. (1976). *Eléments de la théorie des fonctions spéciales*. Moscow: Mir Publishers.
7. Rudin, W. (1966). *Fractional calculus with applications for nuclear reactor dynamics*. New York: McGraw-Hill.
8. von Golitschek, M. (1983). A short proof of Müntz theorem. *Journal of Approximation Theory, 39*, 394–395.

Chapter 6
Numerical Methods

This chapter studies several numerical methods for fractional order systems. In the following sections variational iteration, least squares, Euler's, and Runge–Kutta methods are analyzed.

6.1 Variational Iteration Method for Fractional Differential Equations

In this section we will introduce the variational iteration method (Variational iteration method (VIM)) [3, 4] to solve the nonlinear FDE [2, 5, 6]. We consider the FDE:

$$D^{\alpha} y(t) + R[y(t)] + N[y(t)] = g(t),$$

where $g(t)$ is a given function, $N[y(t)]$ is a nonlinear operator, and R is a residual linear operator. The equation will be solved for the initial conditions:

$$y^{(k)}(0) = y_0^{(k)}, \qquad m - 1 < \alpha \leq \lceil \alpha \rceil = m, \qquad k = 0, \ldots, m - 1,$$

considering that the conditions of existence and uniqueness are satisfied.

We apply the LT method:

$$L[y(t)] = Y(s) = Y,$$

$$L[D^{\alpha} y(t)] = s^{\alpha} Y - \sum_{k=0}^{m-1} y_0^{(k)} s^{\alpha-k-1}.$$

C. Milici et al., *Introduction to Fractional Differential Equations*, Nonlinear Systems and Complexity 25, https://doi.org/10.1007/978-3-030-00895-6_6

We will introduce now the correction relation, written in a recurrent form:

$$Y_{n+1} = Y_n + \lambda(s)\left[s^\alpha Y_n - \sum_{k=0}^{m-1} y_0^{(k)} s^{\alpha-k-1} + L[R[y_n] + N[y_n] - g(t)]\right],$$

where $\lambda(s)$ is the *Lagrange*[1] *multiplier*.

We impose the condition $\dfrac{\delta Y_{n+1}}{\delta Y_n} = 0$. It follows:

$$1 + \lambda(s)s^\alpha = 0 \;\Rightarrow\; \lambda(s) = -\frac{1}{s^\alpha}.$$

It results:

$$Y_{n+1} = \frac{1}{s^\alpha}\sum_{k=0}^{m-1} y_0^{(k)} s^{\alpha-k-1} - \frac{1}{s^\alpha} L[R[y_n] + N[y_n] - g(t)],$$

or:

$$y_{n+1}(t) = L^{-1}\left[\frac{1}{s^\alpha}\sum_{k=0}^{m-1} y_0^{(k)} s^{\alpha-k-1} - \frac{1}{s^\alpha} L[R[y_n] + N[y_n] - g(t)]\right].$$

Example 1 Use the VIM method to solve the FDE:

$$D^\alpha y(t) = 1 + \int_0^t y(u)\,du,$$

with the initial condition:

$$y(0) = 1, \quad 0 < \alpha \leq 1.$$

Solution Using LT we have successively:

$$L[y(t)] = Y,$$

$$s^\alpha Y - s^{\alpha-1} = \frac{1}{s} + L\left[\int_0^t y(u)\,du\right],$$

$$Y_{n+1} = Y_n + \lambda(s)\left[s^\alpha Y_n - s^{\alpha-1} - \frac{1}{s} - L\left[\int_0^t y(u)\,du\right]\right],$$

[1] Joseph-Louis Lagrange (1736–1813).

$$\frac{\delta Y_{n+1}}{\delta Y_n} = 0 \;\Rightarrow\; 1 + \lambda(s)s^{\alpha} = 0 \;\Rightarrow\; \lambda(s) = -\frac{1}{s^{\alpha}},$$

$$y_{n+1} = L^{-1}\left[\frac{1}{s} + \frac{1}{s^{\alpha+1}} + \frac{1}{s^{\alpha}} L\left[\int_0^t y(u)\,du\right]\right],$$

$$y_n(t) = 1 + \frac{t^{\alpha}}{\Gamma(\alpha+1)} + L^{-1}\left[\frac{Y_n}{s^{\alpha+1}}\right]$$

$$y_0 = 1,$$

$$y_1 = 1 + \frac{t^{\alpha}}{\Gamma(\alpha+1)} + \frac{t^{\alpha+1}}{\Gamma(\alpha+2)},$$

$$\cdots$$

Example 2 Use the VIM to solve the FDE:

$$D^{\alpha} y(t) = 1 + \int_0^t (t-u)y(u)\,du,$$

with the conditions:

$$y(0) = 1, \qquad y'(0) = 0, \qquad 1 < \alpha \leq 2.$$

Solution Using LT we have:

$$L[y(t)] = Y,$$

$$s^{\alpha}Y - s^{\alpha-1} = \frac{1}{s} + \frac{1}{s^2} L\left[y_n(u)\right],$$

$$Y_{n+1} = Y_n + \lambda(s)\left[s^{\alpha}Y_n - s^{\alpha-1} - \frac{1}{s} - L\left[y_n\right]\right],$$

$$\frac{\delta Y_{n+1}}{\delta Y_n} = 0 \;\Rightarrow\; 1 + \lambda(s)s^{\alpha} = 0 \;\Rightarrow\; \lambda(s) = -\frac{1}{s^{\alpha}},$$

$$y_{n+1} = L^{-1}\left[\frac{1}{s} + \frac{1}{s^{\alpha+1}} + \frac{1}{s^{\alpha+2}} L\left[y_n\right]\right],$$

$$y_n(t) = 1 + \frac{t^{\alpha}}{\Gamma(\alpha+1)} + L^{-1}\left[\frac{Y_n}{s^{\alpha+2}}\right],$$

$$y_0 = 1,$$

$$y_1 = 1 + \frac{t^\alpha}{\Gamma(\alpha+1)} + \frac{t^{\alpha+2}}{\Gamma(\alpha+3)}.$$

$$\dots$$

Other details on VIM can be found in [2–4].

6.2 The Least Squares Method

Consider the equation

$$D^\alpha y(t) + y(t) + f(t) = 0,$$

with the conditions:

$$y(0) = 0, \qquad y(1) = 0, \qquad 0 < \alpha \le 1,$$

using the approximation:

$$y_{app} = \sum_{i=1}^{n} C_i \phi_i,$$

where C_i are constants, and $\phi_i = \phi_i(t)$, $i = 1, \dots, n$ are test functions. For calculations, we will introduce an operator L:

$$L[y_{app}] = D^\alpha y_{app} + y_{app}.$$

We will define the functional I

$$I[C_1, C_2, \dots, C_n] = \frac{1}{\Gamma(\alpha+1)} \int_0^1 \left[L[y_{app}] + f(t)\right]^2 (dt)^\alpha \to \min,$$

which, by minimization, gives a system of equations in $C_1, \dots, C_n$:

$$\frac{\partial I[C_1, C_2, \dots, C_n]}{\partial C_i} = 0, \qquad i = 1, 2, \dots n.$$

from which we obtain the constants $C_1, \dots, C_n$.

Example 1 Establish an approximate solution, with the aid of least squares method, for the FDE:

$$D^\alpha y(t) + 1 - (1+\alpha)t = 0,$$

with the conditions:

$$y(0) = 0, \qquad y(1) = 0, \qquad 0 < \alpha \leq 1.$$

Solution Exact solution, using the LT method is:

$$s^\alpha Y = -L[1] + (\alpha + 1)L[t], \;\Leftrightarrow\; y(t) = -\frac{t^\alpha}{\Gamma(\alpha + 1)} + \frac{t^{\alpha+1}}{\Gamma(\alpha + 1)}.$$

For

$$y_{app} = C_1\phi_1 + C_2\phi_2,$$

$$D^\alpha y_{app} = C_1 D^\alpha \phi_1 + C_2 D^\alpha \phi_2.$$

we obtain:

$$I[C_1, C_2] = \frac{1}{\Gamma(\alpha + 1)} \int_0^1 \left[C_1 D^\alpha \phi_1 + C_2 D^\alpha \phi_2 + 1 - (\alpha + 1)t\right]^2 (dt)^\alpha \to \min,$$

$$\begin{cases} \dfrac{\partial I[C_1, C_2]}{\partial C_1} = 0, \\ \dfrac{\partial I[C_1, C_2]}{\partial C_2} = 0, \end{cases}$$

$$\begin{cases} \displaystyle\int_0^1 \left[C_1 D^\alpha \phi_1 + C_2 D^\alpha \phi_2 + 1 - (1 + \alpha)t\right] D^\alpha \phi_1 (dt)^\alpha = 0, \\ \displaystyle\int_0^1 \left[C_1 D^\alpha \phi_1 + C_2 D^\alpha \phi_2 + 1 - (1 + \alpha)t\right] D^\alpha \phi_2 (dt)^\alpha = 0, \end{cases}$$

and using the notation

$$B_{ij} = \int_0^1 (D^\alpha \phi_i)(D^\alpha \phi_j)(dt)^\alpha, \quad F_i = -\int_0^1 (1 - (\alpha + 1)t) D^\alpha \phi_i (dt)^\alpha,$$

where $i, j = 1, 2$. It follows:

$$B = \begin{pmatrix} B_{11} & B_{12}, \\ B_{21} & B_{22}, \end{pmatrix}, \qquad C = [C_1, C_2], \qquad F = [F_1, F_2].$$

Finally, the system will be:

$$BC^T = F^T,$$

$$C^T = B^{-1}F^T.$$

Using the formulas:

$$D^\alpha t^k = \frac{\Gamma(k+1)}{\Gamma(k+1-\alpha)} t^{k-\alpha}$$

$$I^\alpha t^k = \frac{\Gamma(k+1)}{\Gamma(k+1+\alpha)} t^{k+\alpha},$$

We choose the test functions as:

$$v_1 = t(1-t), \qquad v_2 = t^2(1-t),$$

$$y_{app} = C_1\, v_1 + C_2\, v_2,$$

$$D^\alpha y_{app} = C_1\, D^\alpha v_1 + C_2\, D^\alpha v_2.$$

$$D^\alpha v_1 = D^\alpha t(1-t) = \frac{1}{\Gamma(2-\alpha)} t^{1-\alpha} - \frac{2}{\Gamma(3-\alpha)} t^{2-\alpha},$$

$$D^\alpha v_2 = D^\alpha (t^2 - t^3) = D^\alpha t^2 - D^\alpha t^3 = \frac{2}{\Gamma(3-\alpha)} t^{2-\alpha} - \frac{6}{\Gamma(4-\alpha)} t^{3-\alpha}.$$

We introduce the notations:

$$P = \frac{1}{\Gamma(2-\alpha)}, \qquad Q = \frac{2}{\Gamma(3-\alpha)}, \qquad R = \frac{6}{\Gamma(4-\alpha)}.$$

We obtain:

$$D^\alpha v_1 = P\, t^{1-\alpha} - Q\, t^{2-\alpha},$$

$$D^\alpha v_2 = P\, t^{2-\alpha} - R\, t^{3-\alpha},$$

$$D^\alpha v_1\, D^\alpha v_1 = P^2\, t^{2-2\alpha} - 2PQ\, t^{3-2\alpha} + Q^2\, t^{4-2\alpha},$$

$$D^\alpha v_1\, D^\alpha v_2 = PQ\, t^{3-2\alpha} - PR\, t^{4-2\alpha} - Q^2\, t^{4-2\alpha} + QR\, t^{5-2\alpha},$$

$$D^\alpha v_2\, D^\alpha v_2 = Q^2\, t^{4-2\alpha} - 2QR\, t^{5-2\alpha} + R^2\, t^{6-2\alpha}.$$

Using the integration rule

$$ {}_0I^{\alpha}x^k = \frac{\Gamma(k+1)}{\Gamma(k+1+\alpha)}, $$

we obtain:

$$ B_{11} = \int_0^1 D^{\alpha}v1D^{\alpha}v1(dt)^{\alpha} = P^2\frac{\Gamma(3-2\alpha)}{\Gamma(3-\alpha)} - 2PQ\frac{\Gamma(4-2\alpha)}{\Gamma(4-\alpha)} + Q^2\frac{\Gamma(5-2\alpha)}{\Gamma(5-\alpha)}, $$

$$ B_{12} = B_{21} = PQ\frac{\Gamma(5-2\alpha)}{\Gamma(5-\alpha)} - PR\frac{\Gamma(5-2\alpha)}{\Gamma(5-\alpha)} - Q^2\frac{\Gamma(5-2\alpha)}{\Gamma(5-\alpha)} + QR\frac{\Gamma(6-2\alpha)}{\Gamma(6-\alpha)}, $$

$$ B_{22} = Q^2\frac{\Gamma(5-2\alpha)}{\Gamma(5-\alpha)} - 2QR\frac{\Gamma(6-2\alpha)}{\Gamma(6-\alpha)} + R^2\frac{\Gamma(7-2\alpha)}{\Gamma(7-\alpha)}, $$

$$ \begin{aligned} F_1 &= -\int_0^1 [1-(\alpha+1)t][Pt^{1-\alpha} - Qt^{2-\alpha}](dt)^{\alpha} \\ &= -P\frac{\Gamma(2-\alpha)}{\Gamma(2)} + Q\frac{\Gamma(3-\alpha)}{\Gamma(3)} + P(\alpha+1)\frac{\Gamma(3-\alpha)}{\Gamma(3)} - Q(\alpha+1)\frac{\Gamma(2-\alpha)}{\Gamma(2)}, \end{aligned} $$

$$ \begin{aligned} F_2 &= -\int_0^1 [1-(\alpha+1)t][Qt^{2-\alpha} - Rt^{3-\alpha}](dt)^{\alpha} \\ &= -Q\frac{\Gamma(3-\alpha)}{\Gamma(3)} + R\frac{\Gamma(4-\alpha)}{\Gamma(4)} + Q(\alpha+1)\frac{\Gamma(4-\alpha)}{\Gamma(4)} - R(\alpha+1)\frac{\Gamma(5-\alpha)}{\Gamma(5)}. \end{aligned} $$

We can build now the matrix B:

$$ B = \begin{pmatrix} B_{11} & B_{12}, \\ & \\ B_{21} & B_{22}, \end{pmatrix}, $$

$$ C = [C_1 \quad C_2], $$

$$ F = [F_1 \quad F_2], $$

and solve the matrix equation:

$$ BC^T = F^T \qquad \Rightarrow \qquad C^T = B^{-1}F^T, $$

resulting C_1 and C_2.

The programs in Maple and Mathematica are listed in the follow-up.

MAPLE

```
> restart:
> Digits:= 4:
> a:= 1/2;
> P:= evalf(1/GAMMA(2 - a));
> Q:= evalf(2/GAMMA(3 - a));
> R:= evalf(6/GAMMA(4 - a));
> B11:=evalf(P^2*GAMMA(3 - 2*a)/GAMMA(3 - a)
  - 2*P*Q*GAMMA(4 - 2*a)/GAMMA(4-a) +
  Q^2*GAMMA(5 - 2*a)/GAMMA(5 - a));
> B12:=evalf(P*Q*GAMMA(4 - 2*a)/GAMMA(4 - a)
  -P*R*GAMMA(5 - 2*a)/GAMMA(5 - a) -  Q^2
  *GAMMA(5 - 2*a)
  /GAMMA(5 - a) + Q*R*GAMMA(6 - 2*a)/GAMMA(6 - a));
> B22:=evalf(P^2*GAMMA(5 - 2*a)/GAMMA(5 - a) -
  2*Q*R*GAMMA(6 - 2*a)/GAMMA(6 - a) + R^2
  *GAMMA(7 - 2*a)
  /GAMMA(7 - a) );
> F1:=evalf(2/GAMMA(3 + a) - 6/GAMMA(4 + a));
> F2:=evalf(6/GAMMA(4 + a) - 24/GAMMA(5 + a));
> ec1:= C1*B11 + C2*B12 = F1;
> ec2:= C1*B12 + C2*B22 = F2;
> solve({ec1,ec2},{C1,C2});
```

MATHEMATICA

```
Clear["`*"]
a = 1/2
P = 1/Gamma[2 - a] // N
Q = 2/Gamma[3 - a] // N
C1 = 6/Gamma[4 - a] // N
v1 = P*t - B*t^2 // N
v2 = P*t^2 - C1*t^3 // N
B11 = P^2*Gamma[3 - 2*a]/Gamma[3 - a] -
   2*P*Q*Gamma[4 - 2*a]/Gamma[4 - a] +
   Q^2*Gamma[5 - 2*a]/Gamma[5 - a] // N
 B12 = P*Q*Gamma[4 - 2*a]/Gamma[4 - a] -
   P*C1*Gamma[5 - 2*a]/Gamma[5 - a] -
   Q^2*Gamma[5 - 2*a]/Gamma[5 - a] +
   Q*C1*Gamma[6 - 2*a]/Gamma[6 - a] // N
B22 = Q^2*Gamma[5 - 2*a]/Gamma[5 - a] -
   2*Q*C1*Gamma[6 - 2*a]/Gamma[6 - a] +
   C1^2*Gamma[7 - 2*a]/Gamma[7 - a] // N
F1 = -P*Gamma[2 - a]/Gamma[2] + Q*Gamma[3 -
a/Gamma[3]] +
```

```
  P*(a + 1)*Gamma[3 - a]/Gamma[3]
-Q*(a + 1)*Gamma[4 - a]/Gamma[4]
F2 = -Q*Gamma[3 - a]/Gamma[3] + C1*Gamma[4 - a]/
  Gamma[4] +
  Q*(a + 1)*Gamma[4 - a]/Gamma[4]
-C1*(a + 1)*Gamma[5 - a]/Gamma[5]
ec1 = C11*B11 + C2*B12
ec2 = C11*B12 + C2*B22
Solve[{ec1, ec2} == {F1, F2}, {C1, C2}] // N
```

For $\alpha = 0.50$, we obtain:

$$C_1 = 54.3332, \qquad C_2 = -39.3793.$$

Example 2 Establish an approximate solution, with the aid of least squares method, for the FDE

$$D^{\alpha} y(t) - y(t) = \frac{t}{e^t - 1}, \qquad 0 \leq t \leq 1,$$

with the initial conditions:

$$y(0) = 1, \qquad y'(0) = 0, \qquad \text{where: } 0 < \alpha \leq 1.$$

Solution It is well known that $f(t)$ can be expanded as:

$$f(t) = \sum_{k=0}^{\infty} B_k \frac{t^k}{k!},$$

where B_k are Bernoulli[2] numbers. In our case:

$$B_0 = 1, \quad B_1 = -\frac{1}{2}, \quad B_2 = \frac{1}{6}, \quad B_4 = -\frac{1}{30}, \ldots, B_{2k+1} = 0.$$

The approximate solution is in this case:

$$y_{app} = \sum_{k=0}^{\infty} C_k t^{\alpha k},$$

where C_k are constants.

[2] J. Bernoulli (1655–1705).

We apply now the least squares method. We introduce the functional:

$$I = \int_0^1 \left[D^\alpha y_{app} - y_{app} - \sum_{k=0}^{\infty} B_k \frac{t^k}{k!} \right]^2 (dt)^\alpha \to \min,$$

which will be minimized. We can calculate also:

$$D^\alpha y_{app} = \sum_{k=0}^{\infty} C_k D^\alpha t^{\alpha k} = \sum_{k=1}^{\infty} C_k \frac{\Gamma(\alpha k + 1)}{\Gamma(\alpha k + 1 - \alpha)} t^{\alpha k - \alpha}$$
$$= \sum_{k=0}^{\infty} C_{k+1} \frac{\Gamma(\alpha k + \alpha + 1)}{\Gamma(\alpha k + 1)} t^{\alpha k},$$

$$I = \int_0^1 \left[\sum_{k=0}^{\infty} C_{k+1} \frac{\Gamma(\alpha k + \alpha + 1)}{\Gamma(\alpha k + 1)} t^{\alpha k} - \sum_{k=0}^{\infty} C_k t^{\alpha k} - \sum_{k=0}^{\infty} B_k \frac{t^k}{k!} \right]^2 (dt)^\alpha \to \min,$$

We apply now the minimization conditions

$$\frac{\partial I}{\partial C_k} = 0 \Rightarrow,$$

from which we obtain the relations:

$$\int_0^1 t^{\alpha k} \left[\sum_{k=0}^{\infty} C_{k+1} \frac{\Gamma(\alpha k + \alpha + 1)}{\Gamma(\alpha k + 1)} t^{\alpha k} - \sum_{k=0}^{\infty} C_k t^{\alpha k} - \sum_{k=0}^{\infty} B_k \frac{t^k}{k!} \right] (dt)^\alpha = 0,$$

where $k = 0, 1, \ldots$

$$\sum_{k=0}^{\infty} \int_0^1 \left[C_{k+1} \frac{\Gamma(\alpha k + \alpha + 1)}{\Gamma(\alpha k + 1)} t^{2\alpha k} - C_k t^{2\alpha k} - \frac{B_k}{k!} t^{k+\alpha k} \right] (dt)^\alpha = 0, \quad k = 0, 1, \ldots$$

We obtain from these relations:

$$C_0 = B_0 = 1,$$

$$C_{k+1} \frac{\Gamma(\alpha k + \alpha + 1)}{\Gamma(\alpha k + 1)} \frac{\Gamma(2\alpha k + 1)}{\Gamma(2\alpha k + \alpha + 1)} + C_k \frac{\Gamma(2\alpha k + 1)}{\Gamma(2\alpha k + \alpha + 1)}$$
$$- B_k \frac{1}{k!} \frac{\Gamma(\alpha k + k + 1)}{\Gamma(\alpha k + k + 1 + \alpha)} = 0,$$

$$C_1 = 0 \qquad C_2 = -\frac{1}{2}\frac{\Gamma(\alpha+2)}{\Gamma(2\alpha+2)}\frac{\Gamma(\alpha+1)\Gamma(3\alpha+1)}{\Gamma^2(2\alpha+1)},$$

$$\dots$$

Finally, we obtain the solution:

$$y_{app}(t) = 1 - \frac{1}{2}\frac{\Gamma(\alpha+2)}{\Gamma(2\alpha+2)}\frac{\Gamma(\alpha+1)\Gamma(3\alpha+1)}{\Gamma^2(2\alpha+1)}t^{2\alpha} + \dots$$

For $\alpha = 1$, we have $y_{app} \approx 1 - \frac{1}{4}t^2$.

Using the Maple sequence of commands:

MAPLE

```
> ec:=diff(y(t),t,t) + y(t) = t/(exp(t) -1 );
> dsolve({ec,y(0) = 1,D(y)0) = 0},y(t),series);
```

it results $y_{app} \approx 1 - \frac{1}{12}t^3$.

Remark Due the fact that FI is difficult to calculate it is recommended that the function $f(t)$ to be represented by a series of powers. So:

$$f(t) = e^t = \sum_{n=0}^{\infty}\frac{t^n}{n!};$$

$$f(t) = \begin{Bmatrix} \cosh(t) \\ \cos(t) \end{Bmatrix} = \sum_{n=0}^{\infty}\frac{(\pm 1)^n}{(2n)!}t^n$$

$$f(t) = \begin{Bmatrix} \sinh(t) \\ \sin(t) \end{Bmatrix} = \sum_{n=0}^{\infty}\frac{(\pm 1)^n}{(2n+1)!}t^n$$

Example 3 Establish the solution, with the aid of least squares method, for the FDE:

$$D^{(3/2)}y(t) - t^{3/2}y(t) = \frac{4t^{1/2}}{\sqrt{\pi}} - t^{7/2},$$

with the initial conditions:

$$y(0) = 0, \qquad y'(0) = 0.$$

Solution Let $B = \{B_0,\ B_1,\ B_2\}$ be basis, with:

$$B_0 = 1,$$
$$B_1 = t,$$
$$B_2 = t^2.$$

We consider a solution of type:

$$y_{ap} = xB_0 + yB_1 + zB_2$$

Using the least squares method we can build the functional I, which will be minimized:

$$I = \frac{1}{\Gamma(5/2)} \int_0^1 \left[D^\alpha y_{app} - t^{3/2} y_{app} - \left(\frac{4t^{1/2}}{\sqrt{\pi}} - t^{3/2} \right) \right]^2 (dt)^{3/2} \to \min,$$

or:

$$I = \frac{1}{\Gamma(5/2)} \int_0^1 \Bigg[(D^{3/2} B_0 - t^{3/2} B_0)\, x + (D^{3/2} B_1 - t^{3/2} B_1)\, y$$
$$+ (D^{3/2} B_2 - t^{3/2} B_2)\, z - \left(\frac{4t^{1/2}}{\sqrt{\pi}} - t^{3/2} \right) \Bigg]^2 (dt)^{3/2} \to \min.$$

We use the notations

$A = D^{3/2} B_0 - t^{3/2} B_0$;
$B = D^{3/2} B_1 - t^{3/2} B_1$;
$C = D^{3/2} B_2 - t^{3/2} B_2$.

The functional I becomes

$$I = \frac{1}{\Gamma(5/2)} \int_0^1 \left[Ax + By + Cz - \left(\frac{4t^{1/2}}{\sqrt{\pi}} - t^{3/2} \right) \right]^2 (dt)^{3/2} \to \min,$$

and the minimization conditions are:

$$\frac{\partial I}{\partial x} = 0, \qquad \frac{\partial I}{\partial y} = 0, \qquad \frac{\partial I}{\partial z} = 0.$$

Explicitly, we have:

$$\begin{cases} \displaystyle\int_0^1 [Ax + By + Cz]\, A\, (dt)^{3/2} = \int_0^1 \left(\frac{4t^{1/2}}{\sqrt{\pi}} - t^{3/2} \right) A (dt)^{3/2}, \\[2ex] \displaystyle\int_0^1 [Ax + By + Cz]\, B\, (dt)^{3/2} = \int_0^1 \left(\frac{4t^{1/2}}{\sqrt{\pi}} - t^{3/2} \right) B\, (dt)^{3/2}, \\[2ex] \displaystyle\int_0^1 [Ax + By + Cz]\, C\, (dt)^{3/2} = \int_0^1 \left(\frac{4t^{1/2}}{\sqrt{\pi}} - t^{3/2} \right) C (dt)^{3/2}. \end{cases}$$

We obtain the system

$$\begin{cases} A_{11}\,x + A_{12}\;y + A_{13}\,z = F_1, \\ A_{21}\,x + A_{22}\;y + A_{23}\,z = F_2, \\ A_{31}\,x + A_{32}\;y + A_{33}\,z = F_3. \end{cases}$$

with:

$$A_{11} = \int_0^1 A^2\,(dt)^{3/2}, \quad A_{12} = \int_0^1 B\,A(dt)^{3/2}, \quad A_{13} = \int_0^1 C\,A\,(dt)^{3/2},$$

$$F_1 = \int_0^1 \left(\frac{4t^{1/2}}{\sqrt{\pi}} - t^{3/2}\right) A\,(dt)^{3/2}$$

$$A_{21} = \int_0^1 A\,B\,(dt)^{3/2}, \quad A_{22} = \int_0^1 B^2(dt)^{3/2}, \quad A_{23} = \int_0^1 C\,B\,(dt)^{3/2},$$

$$F_2 = \int_0^1 \left(\frac{4t^{1/2}}{\sqrt{\pi}} - t^{3/2}\right) B\,(dt)^{3/2}$$

$$A_{31} = \int_0^1 C\,A\,(dt)^{3/2}, \quad A_{32} = \int_0^1 C\,B\,(dt)^{3/2}, \quad A_{13} = \int_0^1 C^2\,(dt)^{3/2},$$

$$F_3 = \int_0^1 \left(\frac{4t^{1/2}}{\sqrt{\pi}} - t^{3/2}\right) C\,(dt)^{3/2}.$$

We observe that:

$$A_{12} = A_{21}, \quad A_{13} = A_{31}, \quad A_{23} = A_{32},$$

$$f(\beta) = \frac{1}{\Gamma(5/2)} \int_0^1 t^{\beta}(dt)^{3/2} = \frac{\Gamma(\beta+1)}{\Gamma(\beta+5/2)}.$$

In **MAPLE**, we obtain exact solution $y(t) = t^2$:

```
> restart;
> Digits:= 6;
> B0:= 1;
> B1:= t;
> B2:=t^2;
> A:= fracdiff(B0,t,3/2) - t^(3/2)*B0;
```

```
> B:= fracdiff(B1,t,3/2) - t^(3/2)*B1;
> C:= fracdiff(B2,t,3/2) - t^(3/2)*B2;
> f:= proc(beta);
> f(beta):=evalf(GAMMA(beta+1)/GAMMA(beta+5/2))
> end proc;
> a11:= collect(A^2,t);
> A11:= f(3);
> a12:= collect(B*A,t);
> A12:= f(4);
> a13:= collect(C*A,t);
> A13:= f(5) - evalf(4/Pi^(1/2))*f(2);
> A21:= A12;
> a22:= collect(B^2,t);
> A22:= f(5);
> a23:= collect(C*B,t);
> A23:= f(6) - evalf(4/sqrt(Pi))*f(3);
> A31:= A13;
> A32:= A23;
> a33:= collect(C^2,t);
> A33:= f(7) - evalf(8/sqrt(Pi))*f(4) + evalf
        (16/Pi)*f(1);
> f1:= collect((4/sqrt(Pi)*t^(1/2) - t^(7/2))*A,t);
> F1:= f(5) - evalf(4/sqrt(Pi))*f(2);
> f2:= collect(((4/sqrt(Pi))*t^(1/2) - t^(7/2))*B,t);
> F2:= f(6) - evalf(4/sqrt(Pi))*f(3);
> f3:= collect((4/sqrt(Pi)*t^(1/2) - t^(7/2))*C,t);
> F3:= f(7) - evalf(8/sqrt(Pi))*f(4)+11/Pi*f(1);
> ec1:= A11*x + A12*y + A13*z = F1;
> ec2:= A21*x + A22*y + A23*z = F2;
> ec3:= A31*x + A32*y + A33*z = F3;
> solve({ec1,ec2,ec3},{x,y,z});
```

The solution is $\{x = 0, \quad y = 0, \quad z = 1\}$, the exact solution being $y(t) = t^2$. In Mathematica we obtain:

MATHEMATICA

```
Clear["Global`*"]
f[x_] := Gamma[x + 1]/Gamma[x + 5/2] // N
B0 = 1
B1 = t
B2 = t^2
A = -t^(3/2)*B0
B = -t^(3/2)*B1
C1 = 4*Sqrt[t] Sqrt[Pi] - t^(3/2)*B2
a11 = A^2
```

```
Expand[%]
A11 = f[3]
a12 = B*A
Expand[%]
A12 = f[4]
a13 = C1*A
Expand[%]
A13 = f[5] - 4/Pi^(1/2)*f[2] // N
A21 = A12
a22 = B^2
Expand[%]
A22 = f[5]
a23 = C*B
Expand[%]
A23 = f[6] - 4/Sqrt[Pi]*f[3] // N
A31 = A13
A32 = A23
a33 = C1^2
Expand[%]
A33 = f[7] - 8/Sqrt[Pi]*f[4] + 16/Pi*f[1] // N
f1 = (4/Sqrt[Pi]*t^(1/2) - t^(7/2))*A
Expand[%]
F1 = f[5] - 4/Sqrt[Pi]*f[2] // N;
f2 = (4/Sqrt[Pi]*t^(1/2) - t^(7/2))*B
Expand[%]
F2 = f[6] - 4/Sqrt[Pi]*f[3] // N
f3 := (4/Sqrt[Pi]*t^(1/2) - t^(7/2))*C
F3 := f[7] - 8/Sqrt[Pi]*f[4] // N
Expand[%]
e3 = f[7] - 8/Sqrt[Pi]*f[4] // N
ec1 = A11*x + A12*y + A13*z
ec2 = A21*x + A22*y + A23*z
ec3 = A31*x + A32*y + A33*z
Solve[{ec1, ec2, ec3} == {F1, F2, F3}, {x, y, z}]
```

6.3 The Galerkin Method for Fractional Differential Equations

The Galerkin[3] method is a direct method for approximate estimation of the solution of FDE. This method will be used here to the FDE:

[3] B.G. Galerkin (1871–1945).

$$D^{\alpha} y(t) + f(t) = 0,$$

where $0 < t < 1$, and $0 < \alpha \leq 1$, and the conditions:

$$y(0) = 0, \qquad y(1) = 0.$$

We denote by $R(t)$ the residual of the equation

$$R(t) = D^{\alpha} y_{app}(t) + f(t),$$

and we will use the approximate solution

$$y_{app}(t) = \sum_{i=1}^{N} C_i \phi_i(t),$$

where $\phi_i(t)$ are the test (or weight) functions, so that

$$\int_0^1 R(t)\phi_i(t)(dt)^{\alpha} = 0, \qquad i = 1, 2, \ldots, N$$

$$\int_0^1 [D^{\alpha} y_{app}(t) + f(t)]\phi_i(t)(dt)^{\alpha} = 0, \qquad i = 1, 2, \ldots, N,$$

$$\int_0^1 D^{\alpha} y_{app}(t)\phi_i(t)(dt)^{\alpha} = -\int_0^1 f(t)\phi_i(t)(dt)^{\alpha}, \qquad i = 1, 2, \ldots, N.$$

Example 1 Use the Galerkin method to find the approximate solution for the FDE:

$$D^{\alpha} y(t) - t = 0,$$

with the conditions:

$$y(0) = 0, \quad y^{(\alpha)}(0) = 0, \qquad 0 < \alpha \leq 1.$$

Solution We choose the test functions:

$$\phi_1 = t, \qquad \phi_2 = t^2,$$

resulting the solution:

$$y_{app} = C_1 t + C_2 t^2.$$

We obtain

$$D^{\alpha} y_{app} = C_1 D^{\alpha} t + C_2 D^{\alpha} t^2.$$

It results the formulas:

$$D^{\alpha} t^k = \frac{\Gamma(k+1)}{\Gamma(k+1-\alpha)} t^{k-\alpha},$$

$$I^{\alpha} t^k = \frac{\Gamma(k+1)}{\Gamma(k+1+\alpha)} t^{k+\alpha},$$

$$D^{\alpha} y_{app} = C_1 \frac{\Gamma(1+1)}{\Gamma(1+1-\alpha)} t^{1-\alpha} + C_2 \frac{\Gamma(2+1)}{\Gamma(2+1-\alpha)} t^{2-\alpha},$$

and finally, we obtain the system

$$\begin{cases} \displaystyle\int_0^1 \left[C_1 \frac{2}{\Gamma(2-\alpha)} t^{1-\alpha} + C_2 \frac{2}{\Gamma(3-\alpha)} t^{2-\alpha} \right] t\,(dt)^{\alpha} = -\int_0^1 t^2 (dt)^{\alpha}, \\ \displaystyle\int_0^1 \left[C_1 \frac{2}{\Gamma(2-\alpha)} t^{1-\alpha} + C_2 \frac{2}{\Gamma(3-\alpha)} t^{2-\alpha} \right] t^2 (dt)^{\alpha} = -\int_0^1 t^3 (dt)^{\alpha}, \end{cases}$$

or:

$$\begin{cases} C_1 \dfrac{2}{\Gamma(2-\alpha)} \dfrac{\Gamma(3-\alpha)}{\Gamma(3)} + C_2 \dfrac{2}{\Gamma(3-\alpha)} \dfrac{\Gamma(4-\alpha)}{\Gamma(4)} = -\dfrac{\Gamma(3)}{\Gamma(3+\alpha)}, \\ C_1 \dfrac{2}{\Gamma(2-\alpha)} \dfrac{\Gamma(4-\alpha)}{\Gamma(4)} + C_2 \dfrac{2}{\Gamma(3-\alpha)} \dfrac{\Gamma(5-\alpha)}{\Gamma(5)} = -\dfrac{\Gamma(4)}{\Gamma(4+\alpha)}. \end{cases}$$

For:
$\alpha = 0.50, \quad \Rightarrow \quad C_1 = -0.4156, \quad C_2 = 0.02536.$
$\alpha = 1, \quad \Rightarrow \quad C_1 = -0.3871, \quad C_2 = 0.08067.$

Example 2 Establish an approximate solution for the following FDE, using the Galerkin method:

$$D^{2\alpha} y(t) + t = 0,$$

with the boundary conditions

$$y(0) = 0, \qquad y(1) = 0, \qquad 0 < \alpha \leq 1.$$

Solution We choose the test functions

$$v_1 = t^2(1-t), \quad v_2 = t^2(1-t^2),$$

and the approximate solution:

$$y_{app} = C_1 v_1 + C_2 v_2.$$

We obtain

$$D^{2\alpha} y_{app} = C_1 D^{2\alpha} v_1 + C_2 D^{2\alpha} v_2,$$

$$D^{2\alpha} t^k = \frac{\Gamma(k+1)}{\Gamma(k+1-2\alpha)} t^{k-2\alpha},$$

and the Galerkin system:

$$\begin{cases} \int_0^1 D^{2\alpha} y_{app} v_1 (dt)^\alpha = 0, \\ \int_0^1 D^{2\alpha} y_{app} v_2 (dt)^\alpha = 0, \end{cases}$$

$$\begin{cases} C_1 \int_0^1 D^{2\alpha} v_1 v_1 (dt)^\alpha + C_2 \int_0^1 D^{2\alpha} v_2 v_1 (dt)^\alpha = -\int_0^1 t v_1 (dt)^\alpha, \\ C_1 \int_0^1 D^{2\alpha} v_1 v_2 (dt)^\alpha + C_2 \int_0^1 D^{2\alpha} v_2 v_2 (dt)^\alpha = -\int_0^1 t v_2 (dt)^\alpha. \end{cases}$$

Introducing the notations

$$A = \frac{2}{\Gamma(3-2\alpha)}, \qquad B = \frac{6}{\Gamma(4-2\alpha)}, \qquad C = \frac{24}{\Gamma(5-2\alpha)},$$

we have:

$$D^{2\alpha} v_1 = D^{2\alpha} t^2 (1-t) = A t^{2-2\alpha} - B t^{3-2\alpha},$$

$$D^{2\alpha} v_2 = D^{2\alpha} (t^2 - t^4) = A t^{2-2\alpha} - C t^{4-2\alpha},$$

$$(D^{2\alpha} v_1) v_1 = A t^{4-2\alpha} - B t^{5-2\alpha} - A t^{5-2\alpha} + B t^{6-2\alpha},$$

$$(D^{2\alpha} v_1) v_2 = A t^{4-2\alpha} - B t^{5-2\alpha} - A t^{6-2\alpha} + B t^{7-2\alpha},$$

$$(D^{2\alpha} v_2) v_1 = A t^{4-2\alpha} - A t^{5-2\alpha} - C t^{6-2\alpha} + C t^{7-2\alpha},$$

$$(D^{2\alpha} v_2) v_2 = A t^{4-2\alpha} - A t^{6-2\alpha} - C t^{6-2\alpha} + C t^{8-2\alpha},$$

$$B_{11} = \int_0^1 (D^{2\alpha} v_1) v_1 (dt)^\alpha$$
$$= A\frac{\Gamma(5-2\alpha)}{\Gamma(5-\alpha)} - B\frac{\Gamma(6-2\alpha)}{\Gamma(6-\alpha)} - A\frac{\Gamma(6-2\alpha)}{\Gamma(6-\alpha)} + B\frac{\Gamma(7-2\alpha)}{\Gamma(7-\alpha)},$$

$$B_{12} = \int_0^1 (D^{2\alpha} v_1) v_2 (dt)^\alpha$$
$$= A\frac{\Gamma(5-2\alpha)}{\Gamma(5-\alpha)} - B\frac{\Gamma(6-2\alpha)}{\Gamma(6-\alpha)} - A\frac{\Gamma(7-2\alpha)}{\Gamma(7-\alpha)} + B\frac{\Gamma(8-2\alpha)}{\Gamma(8-\alpha)},$$

$$B_{21} = \int_0^1 (D^{2\alpha} v_1) v_2 (dt)^\alpha =$$
$$= A\frac{\Gamma(5-2\alpha)}{\Gamma(5-\alpha)} - A\frac{\Gamma(6-2\alpha)}{\Gamma(6-\alpha)} - C\frac{\Gamma(7-2\alpha)}{\Gamma(7-\alpha)} + C\frac{\Gamma(8-2\alpha)}{\Gamma(8-\alpha)},$$

$$B_{22} = \int_0^1 (D^{2\alpha} v_2) v_2 (dt)^\alpha =$$
$$= A\frac{\Gamma(5-2\alpha)}{\Gamma(5-\alpha)} - A\frac{\Gamma(7-2\alpha)}{\Gamma(7-\alpha)} - C\frac{\Gamma(7-2\alpha)}{\Gamma(7-\alpha)} + C\frac{\Gamma(9-2\alpha)}{\Gamma(9-\alpha)},$$

$$F_1 = -\int_0^1 t v_1 (dt)^\alpha = -\frac{\Gamma(4)}{\Gamma(4+\alpha)} + \frac{\Gamma(5)}{\Gamma(5+\alpha)},$$

$$F_2 = -\int_0^1 t v_2 (dt)^\alpha = -\frac{\Gamma(4)}{\Gamma(4+\alpha)} + \frac{\Gamma(6)}{\Gamma(6+\alpha)}.$$

Using the notation

$$B = \begin{pmatrix} B_{11} & B_{12}, \\ B_{21} & B_{22}, \end{pmatrix},$$

$$C = [C_1 \quad C_2],$$

$$F = [F_1 \quad F_2],$$

we obtain in the matrix notation:

$$BC^T = F^T \qquad \Rightarrow \qquad C^T = B^{-1} F^T.$$

Using Maple or Mathematica C_1 and C_2 can be calculated:

MAPLE

```
restart;Digits:=4:
> a:=1/2:
> V1:= t^2 - t^3:
> V2:= t^2 - t^4;
> A:= evalf(2/GAMMA(3 - 2*a));
> B:= evalf(6/GAMMA(4 - 2*a));
> C:= evalf(24/GAMMA(5 - 2*a));
> B11:=evalf(A*GAMMA(5 - 2*a)/GAMMA(5 - a)
>   - B*GAMMA(6 - 2*a)/GAMMA(6 - a)
> - A*GAMMA(6 - 2*a)/GAMMA(6 - a)
> + B*GAMMA(7 - 2*a)/GAMMA(7 - a));
> B12:=evalf(A*GAMMA(5 - 2*a)/GAMMA(5 - a)
>   - B*GAMMA(6 - 2*a)/GAMMA(6 - a)
> - A*GAMMA(7 - 2*a)/GAMMA(7 - a)
> + B*GAMMA(8 - 2*a)/GAMMA(8 - a));
> B21:=evalf(A*GAMMA(5 - 2*a)/GAMMA(5 - a)
> - A*GAMMA(6 - 2*a)/GAMMA(6 - a)
> - C*GAMMA(7 - 2*a)/GAMMA(7 - a)
> + C*GAMMA(8 - 2*a)/GAMMA(8 - a));
> B22:=evalf(A*GAMMA(5 - 2*a)/GAMMA(5 - a)
> - A*GAMMA(7 - 2*a)/GAMMA(7 - a)
> - C*GAMMA(7 - 2*a)/GAMMA(7 - a)
> + C*GAMMA(9 - 2*a)/GAMMA(9 - a));
> F1:=evalf(-GAMMA(4)/GAMMA(4 + a)
  + GAMMA(5)/GAMMA(5 + a));
> F2:=evalf(-GAMMA(4)/GAMMA(4 + a)
  + GAMMA(6)/GAMMA(6 + a));
> ec1:= C1*B11 + C2*B12 = F1;
> ec2:= C1*B21 + C2*B22 = F2;
> solve({ec1,ec2},{C1,C2});
```

MATHEMATICA

```
a= 1/2
A = 2/Gamma[3 - 2*a]//N
B = 2/Gamma[4 - 2*a]//N
C = 24/Gamma[5 - 2*a]//N
B11 = A*Gamma[5 - 2*a]/Gamma[5 - a]
 - B*Gamma[6 - 2*a]/Gamma[6 - a]
 - A*Gamma[6 - 2*a]/Gamma[6 - a]
 + B*Gamma[7 - 2*a]/Gamma[7 - a]//N
B12 = A*Gamma[5 - 2*a]/Gamma[5 - a]
 - B*Gamma[6 - 2*a]/Gamma[6 - a]
```

```
 - A*Gamma[7 - 2*a]/Gamma[7 - a]
 + B*Gamma[8 - 2*a]/Gamma[8 - a]//N
B21 = A*Gamma[5 - 2*a]/Gamma[5 - a]
 - A*Gamma[6 - 2*a]/Gamma[6 - a]
 - C*Gamma[7 - 2*a]/Gamma[7 - a]
 + C*Gamma[8 - 2*a]/Gamma[8 - a]//N
B22 = A*Gamma[5 - 2*a]/Gamma[5 - a]
 - A*Gamma[7 - 2*a]/Gamma[7 - a]
 - C*Gamma[7 - 2*a]/Gamma[7 - a]
 + C*Gamma[9 - 2*a]/Gamma[9 - a]//N
F1 = -Gamma[4]/Gamma[4 + a] + Gamma[5]/
 Gamma[5 + a]//N
F2 = -Gamma[4]/Gamma[4 + a] + Gamma[6]/
 Gamma[6 + a]//N
ec1 = C1*B11 + C2*B12
ec2 = C1*B12 + C2*B22
Solve[{ec1,ec2}=={F1,F2},{F1,F2}]//N
```

For:
$\alpha = 0.50$, $C_1 = -12.89$, $C_2 = 8.135$
$\alpha = 1$, $C_1 = 1.067$, $C_2 = -0.3944$

Example 3 Establish an approximate solution using the Galerkin method, for the FDE:

$$D^{1/2}y(t) = -1 + y^2(t), \qquad y(0) = 0, \qquad 0 < t \leq 1.$$

Solution We apply the LT and Adomian method :

$$L[D^{\alpha}y(t)] = -L[1] + L[y^2(t)], \quad y(0) = 0,$$

(see Example 2 presented in Sec. 4.3.2) solution is:

$$Y_a = -\frac{t^{\alpha}}{\Gamma(\alpha+1)} + \frac{\Gamma(2\alpha+1)}{\Gamma^2(\alpha+1)\Gamma(3\alpha+1)}t^{3\alpha} + \ldots$$

We consider approximate solution

$$Y_a = At^{1/2} + Bt^{3/2},$$

and we have:

$$\begin{cases} \int_0^1 (D^{1/2}Y_a - Y_a^2)t^{1/2}(dt)^{1/2} = -\int_0^1 t^{1/2}(dt)^{1/2}, \\ \int_0^1 (D^{1/2}Y_a - Y_a^2)t^{3/2}(dt)^{1/2} = -\int_0^1 t^{3/2}(dt)^{1/2}. \end{cases}$$

Using Maple or Mathematica A and B can be calculated:

MAPLE

```
> restart:
> Digits:=4:
> a:=1/2:
> Y_a:=A*t^a+B*t^(3*a):
> fracdiff(Y_a,t,1/2):
> unassign('t'):
> g:=t->fracdiff(Y_a,t,1/2)-(Y_a)^2:
> ex1:=g(t)*t^(1/2):
> Ex1:=collect(ex1,t):
> ex2:=g(t)*t^(3/2):
> Ex2:=collect(ex2,t):
> f:=t->GAMMA(t+1)/GAMMA(t+1+1/2):
> f1:=f(1/2):
> f2:=f(3/2):
> ec1:=-B^2*f(7/2)-2*A*B*f(5/2)+(3/4*Pi^(1/2)
     *B-A^2)*f(3/2)+1/2*Pi^(1/2)*A*f(1/2)=-f1:
> ec2:=-B^2*f(9/2)-2*A*B*f(7/2)+(3/4*Pi^(1/2)
     *B-A^2)*f(5/2)+1/2*Pi^(1/2)*A*f(3/2)=-f2:
> ecu1:=evalf(ec1):
> ecu2:=evalf(ec2):
> solve({ecu1,ecu2},{A,B});
  {A = -0.9740, B = 0.2786}, {A = -5.798, B = 9.315},
  {A = 4.568 - 0.6385 I, B = -3.956 + 1.690 I},
  {A = 4.568 + 0.6385 I, B = -3.956 - 1.690 I}
```

MATHEMATICA

```
Clear["`*"]
f[x_] := Gamma[x + 1]/Gamma[x + 3/2];
Ya = A*t^(1/2) + B*t^(3/2);
DYa = A*Sqrt[Pi]/2 + B*3*Sqrt[Pi]/4*t;
g[t_] := DYa - (Ya)^2;
ex1 = g[t]*t^(1/2);
Ex1 = Expand[%];
ex2 = g[t]*t^(3/2);
Ex2 = Expand[%];
f1 = -f[1/2] // N;
f2 = -f[3/2] // N;
ex1 = (Sqrt[Pi]/2*A + 3*Sqrt[Pi]/4*B - (A*Sqrt[t]+
      B*t^(3/2))^2)*Sqrt[t];
Ex1 = Expand[%];
ex2 = (Sqrt[Pi]/2*A + 3*Sqrt[Pi]/4*B - (A*Sqrt[t]+
      B*t^(3/2))^2)*t^(3/2);
```

```
Ex22 = Expand[%];
ec1 = 1/2*A*Sqrt[Pi]*f[1/2] + 3/4*B*Sqrt[Pi]*f[3/2]
    - A^2*f[3/2] - 2 A*B*f[5/2] - B^2*f[7/2] // N;
ec2 = 1/2*A*Sqrt[Pi]*f[3/2] + 3/4*B*Sqrt[Pi]*f[5/2]
    - A^2*f[5/2] - 2 A*B*f[7/2] - B^2*f[9/2] // N;
NSolve[{ec1, ec2} == {f1, f2}, {A, B}]
```

6.4 Euler's Method

We introduce here a generalization of the Euler method to the case of FDE of type:

$$D^{\alpha} y(t) = f(t, y(t)), \qquad \text{with: } y(t_0) = y_0, \quad \text{where: } 0 < \alpha \leq 1.$$

We assume that $y(t)$, $D^{\alpha} y(t)$, $D^{2\alpha} y(t)$ are continuous on $[t_0, a]$ and we will apply the generalized Taylor's formula:

$$y(t) = y(t_0) + D^{\alpha} y(t_0) \frac{t^{\alpha}}{\Gamma(\alpha+1)} + D^{2\alpha}(\eta) \frac{t^{2\alpha}}{\Gamma(2\alpha+1)}, \qquad t_0 < \eta \leq t.$$

We obtain the iterative Euler's formula:

$$y(t) \approx y(t_0) + \frac{h^{\alpha}}{\Gamma(\alpha+1)} f(t_0, y(t_0)),$$

or, expressed in the recurrent form:

$$y_{n+1} \approx y_n + \frac{h^{\alpha}}{\Gamma(\alpha+1)} f(t_n, y_n),$$

Example 1 Solve the following FDE:

$$D^{1/2} y(t) = y(t) - \frac{2t}{y(t)}, \qquad \text{with the initial condition: } y(0) = 1,$$

where $t \in [0, 1]$, and step $h = 0.2$.

Solution We can calculate the solution in Maple and Mathematica using the Euler formula:

MAPLE

```
> restart;
> Digits:=4:
> unassign(t,y);
> f:=(t,y)->y-2*t/y:
```

```
> d:=0.2:
> a:=1/2:
> h:=evalf(d^a/GAMMA(a+1)):
> t:=array[0..10]:
> y:=array[0..10]:
> for k from 0 to 5 do t[k]:=k*0.2 od:
> y[0]:=1:
> for k from 0 to 5 do y[k+1]:=y[k]+h*f(t[k],y[k]) od:
> for k from 0 to 5 do print(t[k]," " ,y[k]) od;
0.0, 1
0.2, 1.504
0.4, 2.128
0.6, 3.012
0.8, 4.331
1.0, 6.329
```

MATHEMATICA

```
Clear["*"]
\[Delta] = 0.1; \[Alpha] = 1/2;
f[t_, y_] := y - 2*t/y
values =
RecurrenceTable[{t[k + 1] == t[k] + \[Delta],
y[k + 1] ==
y[k] + \[Delta]^\[Alpha]/Gamma[\[Alpha] + 1]
  *f[t[k], y[k]],
t[0] == 0, y[0] == 1}, {t, y}, {k, 0, 10}];
  Grid[values]
ListLinePlot[values, PlotMarkers -> Automatic]
```

The solution $y(t)$ is plotted in Fig. 6.1.

Example 2 Approximate the following FDE with the aid of Euler method:

$$D^{\alpha}y(t) = \frac{y(t) - t}{y(t) + t}, \quad \text{for the initial condition: } y(0) = 1,$$

with the step $h = 0.1$, and $t \in [0, 1]$.

MAPLE

```
< restart;
> Digits:=4:
> unassign(t,y);
> f:=(t,y)->(y- t)/(y+t):
> d:=0.1:
> a:=1/2:
> h:=evalf(d^a/GAMMA(a+1)):
```

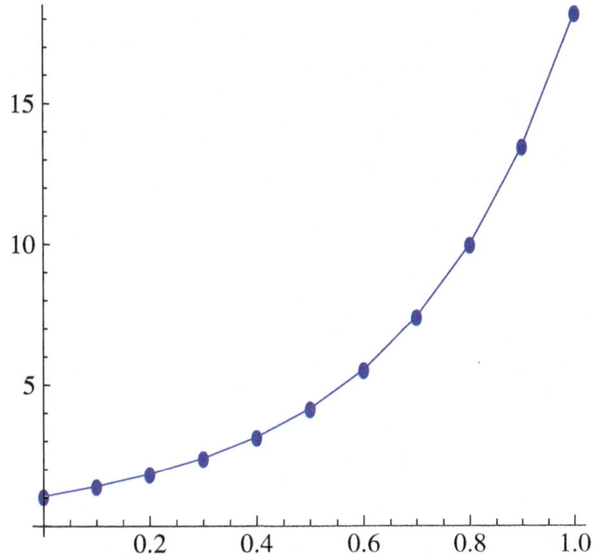

Fig. 6.1 The solution $y(t)$ of Example 1

```
> t:=array[0..10]:
> y:=array[0..10]:
> for k from 0 to 10 do t[k]:=k*0.1 od:
> y[0]:=1:
> for k from 0 to 10 do y[k+1]:=y[k]+h*f(t[k],
  y[k]) od:
> for k from 0 to 10 do print(t[k]," " ,y[k]) od;
0., 1
0.1, 1.357
0.2, 1.665
0.3, 1.945
0.4, 2.206
0.5, 2.453
0.6, 2.689
0.7, 2.916
0.8, 3.135
0.9, 3.347
1.0, 3.552
```

MATHEMATICA

```
Clear["*"]
\[Delta] = 0.1; \[Alpha] = 1/2;
f[t_, y_] := (y - t)/(y + t)
values =
RecurrenceTable[{t[k + 1] == t[k] + \[Delta],
```

```
y[k + 1] ==
y[k] + \[Delta]^\[Alpha]/Gamma[\[Alpha] + 1]
  *f[t[k], y[k]],
t[0] == 0, y[0] == 1}, {t, y}, {k, 0, 10}];
  Grid[values]
ListLinePlot[values, PlotMarkers -> Automatic]
0.0 1
0.1 1.35582
0.2 1.86466
0.3 1.94494
0.4 2.2064
0.5 2.4532
0.6 2.68972
0.7 2.91639
0.8 3.13587
0.9 3.34681
1 3.5524
```

In this case the solution $y(t)$ is plotted in Fig. 6.2.

Using the **eulerstep[...]** command, the solution can be obtained using the Mathematica program:

MATHEMATICA

```
Clear["`*"]
a = 1/2;
d = 0.1;
eulerstep[f_, {t_, y_}, h_] := {t + h,
```

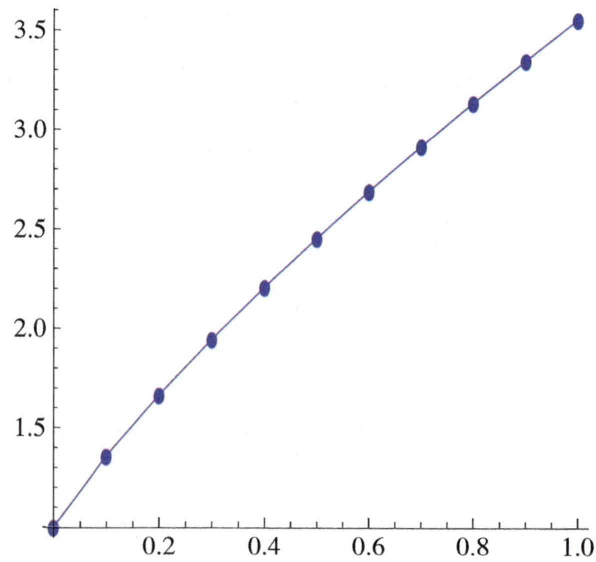

Fig. 6.2 The solution $y(t)$ of Example 2

```
    y + d ^a/Gamma[a + 1] f[t, y]}
    euler[f_, {t_, y_}, tf_, h_] :=
    NestList[eulerstep[f, #1, h] &, {t, y},
    Ceiling[(tf - t)/h]]
   f[t_, y_] := (y - t)/(y + t)
   tmp = euler[f, {0, 1}, 1, 0.1];
   PaddedForm[TableForm[tmp], {6, 4}]
     0.0,    1.0,
     0.1,    1.3568,
     0.2,    1.6646,
     0.3,    1.9449,
     0.4,    2.2064,
     0.5,    2.4537,
     0.6,    2.6897,
     0.7,    2.9163,
```

Example 3 Approximate the system of FDE:

$$\begin{cases} D^{1/2}x(t) = \dfrac{y(t) - t}{t}, \\ D^{1/2}y(t) = \dfrac{x(t) + t}{t}, \end{cases}$$

with the initial conditions $x(1) = 1,\ y(1) = 1$, the time step $h = 0.2$ and $t \in [1, 2]$.

Solution We can use the following program:

MAPLE

```
> restart; Digits:=4:
> unassign(t,x,y);
> f:=(t,x,y)->(y-x)/t:g:=(t,x,y)->(y+x)/t:
> d:=0.2:
> a:=1/2:
> h:=evalf(d^a/GAMMA(a+1)):
> t:=array[1..10]:x:=array[1..10]:
> y:=array[1..10]:t[1]:=1:
> for k from 1 to 6 do t[k+1]:=t[k]+0.2 od:
> y[1]:=1:x[1]:=1:
> for k from 1 to 6 do
           x[k+1]:=x[k]+h*f(t[k],x[k],y[k]):
           y[k+1]:=y[k]+h*g(t[k],x[k],y[k]) od;
> for k from 1 to 6 do print(t[k],x[k] ,y[k]) od;
1,0  1,      1
1.2, 1.,     2.009
1.4, 1.424, 3.274
1.6, 2.090, 4.967
```

```
1.8, 2.997, 7.192
2.0, 4.173, 10.05
```

MATHEMATICA

```
Clear["*"]
\[Delta] = 0.2; \[Alpha] = 1/2;
f[t_, x_, y_] := (y - x)/t
g[t_, x_, y_] := (y + x)/t
values = RecurrenceTable[{t[k + 1] == t[k] + \[Delta],
x[k + 1] == x[k] + \[Delta]^\[Alpha]/Gamma[\[Alpha]
                                + 1]*f[t[k], x[k], y[k]],
y[k + 1] == y[k] + \[Delta]^\[Alpha]/Gamma[\[Alpha]
                                + 1]*g[t[k], x[k], y[k]],
t[0] == 1, x[0] == 1, y[0] == 1}, {t, x, y},
  {k, 0, 5}]//
TableForm
1   1       1
1.2 1       2.00925
1.4 1.42441 3.27471
1.6 2.09135 4.96135
1,8 2.91878 7.19151
2   4.17521 10.053
```

Remark Using the notations

$$k_1 = f(t_n, y_n),$$

$$k_2 = f\left(t_n + \frac{h}{2}, y_n + \frac{k_1}{2}\right),$$

a more general Euler's iterative formula can be written as:

$$y_{n+1} = y_n + h\,k_2.$$

Example 4 (Lorenz Attractor) Solve the Lorentz attractor system:

$$\begin{cases} D^{0.98}x(t) = -10(x - y), & x(0) = 0, \\ D^{0.98}y(t) = 28x - y - xz, & y(0) = 1, \\ D^{0.98}z(t) = xy - \frac{8}{3}z, & z(0) = 0. \end{cases}$$

The solution in Mathematica is:

```
Clear["`*"]
d = 0.01
```

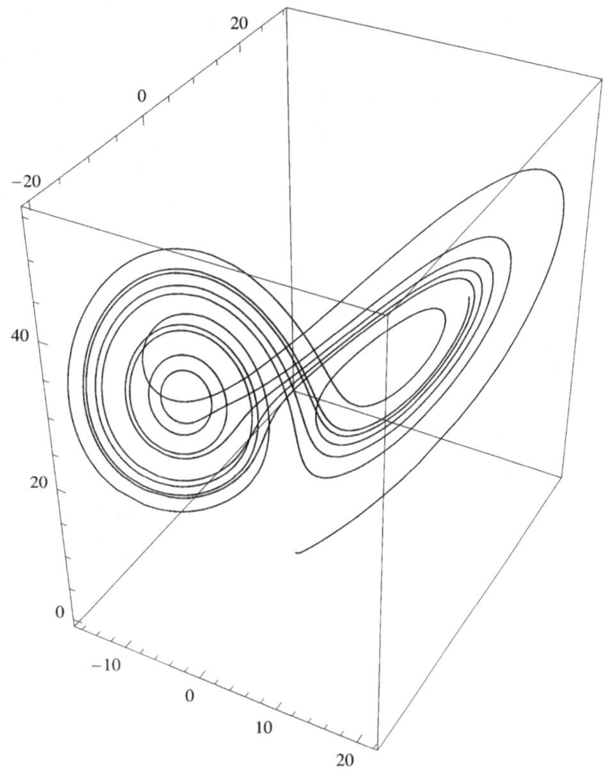

Fig. 6.3 The 3D Lorenz attractor solution $(x(t), y(t), z(t))$ for $\alpha = 0.98$

```
a = 0.98
h = d^a/Gamma[a + 1]
Eulerlor[{x_, y_, z_}] := {x - 10*h*(x - y),
                     y + h*(28*x - y - x*z),
                     z + h*(x*y - 8/3*z)}
sol = NestList[Eulerlor, {0, 1, 0}, 1000];
p = Interpolation /@ Transpose@sol;
ParametricPlot3D[Through@p@t, {t, 1, 1000},
     PlotPoints -> 100,
 ColorFunction -> (Hue[4 #] &), ImageSize -> 300]
```

Figure 6.3 shows the 3D Lorenz attractor solution $(x(t), y(t), z(t))$ for $\alpha = 0.98$.

6.5 Runge–Kutta Methods for Fractional Differential Equation

In this section we will suppose that all conditions of existence and uniqueness of the solutions are fulfilled.

6.5.1 The Second Order Runge–Kutta Method

The second order Runge[4]–Kutta[5] method Second order Runge–Kutta (RK2) introduced here is an extension of the first Euler method for approximation of the solution of the FDE:

$$D^{\alpha} y(t) = f(t, y(t)), \qquad y(t_0) = y_0, \qquad 0 < \alpha \leq 1,$$

where:

$$y \in C^{p+1}([t_0, \ t_0 + T]).$$

In this case the solution can be approximated in the following discrete form:

$$y_{n+1} = y_n + \frac{h^{\alpha}}{\Gamma(\alpha+1)} \frac{K_1 + K_2}{2},$$

where

$$K_1 = f(t_n, y_n),$$

$$K_2 = f\left(t_n + \frac{h^{\alpha}}{\Gamma(\alpha+1}, y_n + \frac{h^{\alpha}}{\Gamma(\alpha+1)} K_1\right).$$

Proof We introduce the notations

$$K_1 = f(t_n, y_n),$$

$$K_2 = f\left(t_n + A\frac{h^{\alpha}}{\Gamma(\alpha+1}, y_n + B\frac{h^{\alpha}}{\Gamma(\alpha+1} K_1\right),$$

where A and B are two unknown real constants.

The local error from the Taylor expansion is

$$E(h) = y_{n+1} - y_n,$$

with:

$$y_{n+1} = y(t_n + h), \qquad t_{n+1} = t_n + h, \qquad y_n = y(t_n).$$

---

[4]C.D.T. Runge (1856–1927).

[5]M.W. Kutta (1867–1944).

Obviously, for minimization of error, we impose the conditions:

$$E(0) = D^{\alpha}E(0) = \ldots = D^{p\alpha}E(0) = 0, \qquad D^{(p+1)\alpha} \neq 0.$$

Using the Taylor polynomial we obtain:

$$\begin{aligned} E(h) = y_{n+1} - y_n &= \frac{h^{\alpha}}{\Gamma(\alpha+1)} D^{\alpha} y_n + \frac{h^{2\alpha}}{\Gamma(2\alpha+1)} D^{2\alpha} y_n + \mathcal{O}(h^{3\alpha}) \\ &= \frac{h^{\alpha}}{\Gamma(\alpha+1)} f(t_n, y_n) + \frac{h^{2\alpha}}{\Gamma(2\alpha+1)} D^{2\alpha} y_n + \mathcal{O}(h^{3\alpha}), \end{aligned}$$

$$D^{\alpha}E(h) = f(t_n, y_n) + \frac{h^{\alpha}}{\Gamma(\alpha+1)} D^{\alpha} y_n + \mathcal{O}(h^{2\alpha}),$$

and

$$y_{n+1} - y_n = C_1 \frac{h^{\alpha}}{\Gamma(\alpha+1)} K_1 + C_2 \frac{h^{\alpha}}{\Gamma(\alpha+1)} K_2.$$

Hence, it results $E[h]$ and $D^{\alpha}E[h]$:

$$E[h] = C_1 \frac{h^{\alpha}}{\Gamma(\alpha+1)} K_1 + C_2 \frac{h^{\alpha}}{\Gamma(\alpha+1)} K_2,$$

$$D^{\alpha}E[h] = c_1 K_1 + c_2 K_2 + \frac{h^{\alpha}}{\Gamma(\alpha+1)} D^{\alpha} K_2,$$

$$\begin{aligned} D^{\alpha} K_2 &= D^{\alpha} f\left(t_n + A\frac{h^{\alpha}}{\Gamma(\alpha+1}, y_n + B\frac{h^{\alpha}}{\Gamma(\alpha+1} K_1\right) \\ &= f_{t_n + A\frac{h^{\alpha}}{\Gamma(\alpha+1)}} A + f_{y_n + B\frac{h^{\alpha}}{\Gamma(\alpha+1)} K_1} B K_1. \end{aligned}$$

For $h \to 0$ we obtain:

$$D^{\alpha}E[0] = C_1 K_1[0] + C_2 K_2[0],$$

$$f(t_n, y_n) = C_1 f(t_n, y_n) + C_2 f(t_n, y_n), \;\Rightarrow\; 1 = C_1 + C_2,$$

$$D^{2\alpha}E(h) = D^{\alpha}(D^{\alpha}E(h)) = C_2 D^{\alpha} K_2 + D^{\alpha} K_2.$$

But $D^{2\alpha}E(0) = D^{2\alpha} y_n(t)$, and:

$$D^{2\alpha} y(t) = D^{\alpha}(f(t, y(t)) = f_t + f_y D^{\alpha} y(t) = f_t + f f_y.$$

It results:

$$2C_2Af_{t_n}(t_n, y_n) + 2C_2Bf(t_n, y_n)f_{y_n}(t_n, y_n) = f_{t_n}(t_n, y_n) + f(t_n, y_n)f_{y_n}(t_n, y_n).$$

We obtain finally the system:

$$\begin{cases} C_1 + C_2 = 1 \\ 2C_2A - 1 = 0 \\ 2C_2B - 1 = 0 \end{cases}$$

For $C_1 = C_2 = \dfrac{1}{2}$ we have the *classical* RK2 method, with the step $\dfrac{h^\alpha}{\Gamma(\alpha+1)}$.

Example Solve with RK2 the FDE

$$D^{1/2}y(t) = t^2 + \frac{y^2}{4}, \qquad y(0) = 1.$$

The solution can be written in Maple or Mathematica:

MAPLE

```
> restart;
> Digits:=3:
> a:=1/2:
> d:=0.2:
> h:=evalf(d^a/GAMMA(a+1)):
> y:=Array[0..10]:
> y[0]:=1:
> unassign('t,y');
> f:=(t,y)->t^2+y^2/4:
> for i from 0 to 4  do
> k1:=f(i*0.2,y[i]):
> k2:=f(i*0.2 + h,y[i]+h*k1):
> y[i+1]:=y[i]+(k1+k2)/2 od:
> for i from 0 to 5  do print(i*0.2,y[i]) od;
  0      1
  0.2    1.207
  0.4    1.560
  0.6    2.200
  0.8    3.468
  1      6.597
```

MATHEMATICA

```
Clear["`*"]
a = 1/2
d = 0.2
h = d^a/Gamma[a + 1]
y[0] = 1
f[t_, y_] := t^2 + y^2/4
For[i = 0, i < 5,
 i++, {k1 = f[t, y] /. {t -> i*0.2, y -> y[i]},
  k2 = f[t, y] /. {t -> i*0.2 + h, y -> y[i] + h*k1},
  y[i + 1] = y[i] + (k1 + k2)*0.5*h}]
Table[{j*0.2, y[j]}, {j, 0, 5}] // TableForm
  0      1
  0.2    1.20733
  0.4    1.56029
  0.6    2.20011
  0.8    3.469
  1      6.59765
```

6.5.2 The Fourth Order Runge–Kutta Method

In the case of fourth order Runge–Kutta Fourth order Runge–Kutta (RK4) algorithm, we will establish an algorithm to solve the following FDE:

$$D^{\alpha} y(t) = f(t, y(t)), \quad y(t_0) = y_0, \quad 0 < \alpha \leq 1,$$

where:

$$y \in C^{p+1}([t_0, \quad t_0 + T]).$$

In the neighborhood of t_0 we suppose that

$$D^{\alpha} E(0) = D^{2\alpha} E(0) = 0, \quad E(h) = y(t + h) - y(t).$$

The approximate solution can be obtained from the expansion

$$y_{n+1} = y_n + \frac{h^{\alpha}}{6\Gamma(\alpha + 1)}(K_1 + 2K_2 + 2K_3 + K_4),$$

where $K_1, \ldots, K_4$ are functions which will be established in the proof.

Proof The proof is similar to the previous case (RK2).

We introduce here the expansion error:

$$\begin{aligned} E(h) = y_{n+1} - y_n &= \frac{h^\alpha}{\Gamma(\alpha+1)} D^\alpha y_n + \frac{h^{2\alpha}}{\Gamma(2\alpha+1)} D^{2\alpha} y_n + \mathcal{O}(h^{3\alpha}) \\ &= \frac{h^\alpha}{\Gamma(\alpha+1)} f(t_n, y_n) + \frac{h^{2\alpha}}{\Gamma(2\alpha+1)} D^{2\alpha} y_n + \mathcal{O}(h^{3\alpha}), \end{aligned}$$

and its derivative of order α:

$$D^\alpha E(h) = f(t_n, y_n) + \frac{h^\alpha}{\Gamma(\alpha+1)} D^\alpha y_n + \mathcal{O}(h^{2\alpha}),$$

and

$$y_{n+1} - y_n = C_1 \frac{h^\alpha}{\Gamma(\alpha+1)} K_1 + C_2 \frac{h^\alpha}{\Gamma(\alpha+1)} K_2 + C_3 \frac{h^\alpha}{\Gamma(\alpha+1)} K_3 + C_4 \frac{h^\alpha}{\Gamma(\alpha+1)} K_4.$$

Hence, we have

$$E[h] = C_1 \frac{h^\alpha}{\Gamma(\alpha+1)} K_1 + C_2 \frac{h^\alpha}{\Gamma(\alpha+1)} K_2 + C_3 \frac{h^\alpha}{\Gamma(\alpha+1)} K_3 + C_4 \frac{h^\alpha}{\Gamma(\alpha+1)} K_4,$$

where:

$$K_1 = f(t_n, y_n),$$

$$K_2 = f\left(t_n + a_2 \frac{h^\alpha}{2\Gamma(\alpha+1)}, \quad y_n + b_2 \frac{h^\alpha}{2\Gamma(\alpha+1)} K_1\right),$$

$$K_3 = f\left(t_n + a_3 \frac{h^\alpha}{2\Gamma(\alpha+1)}, \quad y_n + b_3 \frac{h^\alpha}{2\Gamma(\alpha+1)} K_2\right),$$

$$K_4 = f\left(t_n + a_4 \frac{h^\alpha}{\Gamma(\alpha+1)}, \quad y_n + b_4 \frac{h^\alpha}{\Gamma(\alpha+1)} K_3\right),$$

and finally:

$$\begin{aligned} D^\alpha E[h] = {} & C_1 K_1 + C_2 K_2 + C_3 K_3 + C_4 K_4 \\ & + C_1 \frac{h^\alpha}{\Gamma(\alpha+1)} D^\alpha K_1 + C_2 \frac{h^\alpha}{\Gamma(\alpha+1)} D^\alpha K_2 \\ & + C_3 \frac{h^\alpha}{\Gamma(\alpha+1)} D^\alpha K_3 + C_4 \frac{h^\alpha}{\Gamma(\alpha+1)} D^\alpha K_4. \end{aligned}$$

We take now $h \to 0$:

$$D^{\alpha} E[0] = C_1 K_1[0] + C_2 K_2[0] + C_3 K_3[0] + C_4 K_4[0]$$

$$f(t_n, y_n) = (C_1 + C_2 + C_3 + C_4) f(t_n, y_n).$$

Hence, we obtain the equation:

$$1 = C_1 + C_2 + C_3 + C_4.$$

From the other conditions we get:

$$\begin{cases} C_2 a_2 + C_3 a_3 + C_4 a_4 = \frac{1}{2}, \\ C_2 b_2 + C_3 b_3 + C_4 b_4 = \frac{1}{2}, \end{cases}$$

(i.e., from $D^{\alpha} E(h) = 0$).

Finally, we have the system:

$$\begin{cases} C_1 + C_2 + C_3 + C_4 = 1, \\ C_2 a_2 + C_3 a_3 + C_4 a_4 = \frac{1}{2}, \\ C_2 b_2 + C_3 b_3 + C_4 b_4 = \frac{1}{2}. \end{cases}$$

Remark The solution of this system is not unique. For this reason we choose as in the RK2 case:

$$C_1 = C_4 = 1/6, \qquad C_2 = C_4 = 1/3$$

$$a_1 = b_1 = a_4 = b_4 = 1, \qquad a_2 = b_2 = a_3 = b_3 = \frac{1}{2}$$

Example 1 Solve, using the RK4 algorithm, the following FDE:

$$D^{\alpha} y(t) = t^2 + \frac{y^2}{4}, \qquad y(0) = 1,$$

for the cases:

(a) $\alpha = a = \frac{1}{2}$,
(b) $\alpha = a = 0.98$.

The solution can be written as:

MATHEMATICA

```
Clear["`*"];
a = 1/2;
y[0] = 1;
h = (0.01)^a/Gamma[a + 1];
n = 5;
f[t_, y_] = t^2 + y^2/4;
Do[{K1 = f[i h, y[i]], K2 = f[i*h + h/2, y[i]
  + h/2*K1],
  K3 = f[i h + h/2, y[i]+h*K2/2], K4 =f[i h + h,
  y[i]+h*K3],
  y[i + 1] = y[i] + (K1 + 2 K2 + 2 K3 + K4)*h/6},
  {i, 0, n}]
Do[Print[PaddedForm[i h, {6, 4}], "          ",
   PaddedForm[y[i], {6, 4}]], {i, 0, 5}];
  0.0000         1.0000
  0.1128         1.0295
  0.2257         1.0637
  0.3385         1.1060
  0.4514         1.1599
  0.5642         1.2292

Clear["`*"]
RK4step[{t_, y_}] := Module[{k1, k2, k3, k4},
  k1 = f[t, y];
  k2 = f[t + h/2, y + h/2 k1];
  k3 = f[t + h/2, y + h/2 k2];
  k4 = f[t + h, y + h  k3];
  {t + h, y + 1/6 (k1 + 2 k2 + 2 k3 + k4)}]
f[t_, y_] := t^2 + y^2/4;
t0 = 0;
y0 = 1;
a = 1/2;
h = (0.01)^a/Gamma[a + 1];
n = 5;
rkpoints = NestList[RK4step, {t0, y0}, n];
PaddedForm[TableForm[rkpoints], {6, 4}]
   0,        1,
   0.1128,   1.2616,
   0.2256,   1.7047,
   0.3385,   2.5522,
   0.4513,   4.4782,
   0.5641,   10.500
```

Example 2 Solve the following system of FDE:

$$\begin{cases} D^{1/2}x(t) = t^2 + \dfrac{y^2}{4}, & x(0) = 0, \\ \\ D^{1/2}y(t) = t^2 + \dfrac{x^2}{4}, & y(0) = 1. \end{cases}$$

Solution

MATHEMATICA

```
Clear["`*"];
a = 1/2; x[0] := 0;
y[0] = 1; h = (0.01)^a/Gamma[a + 1]; n = 5;
Do[t[i] = 0.0 + (0.2)*i, {i, 0, n}]
f[t_, x_, y_] = t^2 + y^2/4
g[t_, x_, y_] = t^2 + x^2/4
Do[{K1 = f[t[i], x[i], y[i]],
  L1 = g[t[i], x[i], y[i]],
  K2 = f[t[i] + h, x[i] + h*K1, y[i] + h*L1],
  L2 = g[t[i] + h, x[i] + h*K1, y[i] + h*L1],
  K3 = f[t[i] + h/2, x[i] + h*K2/2, y[i] + h*L2/2],
  L3 = g[t[i] + h/2, x[i] + h*K2/2, y[i] + h*L2/2],
  K4 = f[t[i] + h, x[i] + h*K3, y[i] + h*L3],
  L4 = g[t[i] + h, x[i] + h*K3, y[i] + h*L3],
  x[i + 1] = x[i] + (K1 + 2 K2 + 2 K3 + K4)*h/6,
  y[i + 1] = y[i] + (L1 + 2 L2 + 2 L3 + L4)*h/6},
  {i, 0, n}]
Do[Print[t[i], "    ", x[i], "   ", y[i]], {i, 0, 5}];
 0     0         1
 0.2   0.0290    0.0088
 0.4   0.0663    1.0096
 0.6   0.1216    1.0356
 0.8   0.3054    1.0881
 1     0.3189    1.1771
```

Example 3 Solve the fractional Van der Pol[6] system:

$$\begin{cases} D^{1/2}x(t) = y, \qquad x(0) = 1, \\ \\ D^{1/2}y(t) = -x + 0.25\,(1 - x^2)\,y, \qquad y(0) = 0. \end{cases}$$

[6]B. van der Pol (1889–1959).

Solution

MATHEMATICA

```
Clear["`*"];
a = 1/2; x[0] := 1;
y[0] = 0; h = (0.2)^a/Gamma[a + 1]; n = 10000;
Do[t[i] = 0.0 + (0.2)*i, {i, 0, n}]
f[t_, x_, y_] = y
g[t_, x_, y_] = -x + 0.25*(1 - x^2)*y
Do[{K1 = f[t[i], x[i], y[i]],
   L1 = g[t[i], x[i], y[i]],
   K2 = f[t[i] + h, x[i] + h*K1, y[i] + h*L1],
   L2 = g[t[i] + h, x[i] + h*K1, y[i] + h*L1],
   K3 = f[t[i] + h/2, x[i] + h*K2/2, y[i] + h*L2/2],
   L3 = g[t[i] + h/2, x[i] + h*K2/2, y[i] + h*L2/2],
   K4 = f[t[i] + h, x[i] + h*K3, y[i] + h*L3],
   L4 = g[t[i] + h, x[i] + h*K3, y[i] + h*L3],
   x[i + 1] = x[i] + (K1 + 2 K2 + 2 K3 + K4)*h/6,
   y[i + 1] = y[i] + (L1 + 2 L2 + 2 L3 + L4)*h/6},
   {i, 0, n}];
Do[Print[t[i], "   ", x[i], "  ", y[i]], {i, 0, 5}];
```

Figure 6.4 shows the plot of the Van der Pol solution from Example 3.

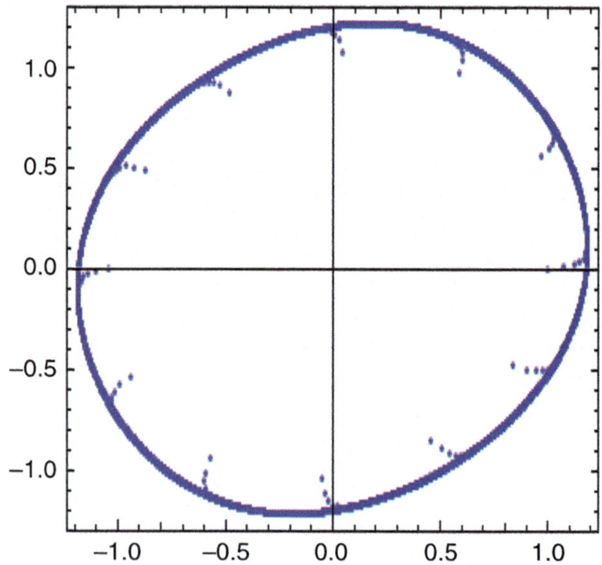

Fig. 6.4 Plot of the Van der Pol solution from Example 3

Example 4 Solve the fractional Duffing[7] system:

$$\begin{cases} D^{0.998}x(t) = y, & x(0) = 1, \\ D^{0.998}y(t) = -x - x^3, & y(0) = 0. \end{cases}$$

Solution

```
Clear["`*"];
a = 0.998; x[0] = 1;
y[0] = 0; h = (0.2)^a/Gamma[a + 1]; n = 10000;
Do[t[i] = 0.0 + (0.2)*i, {i, 0, n}]
f[t_, x_, y_] := y
g[t_, x_, y_] := -x - x^3
Do[{K1 = f[t[i], x[i], y[i]], L1 = g[t[i], x[i],
  y[i]],
  K2 = f[t[i] + h, x[i] + h*K1, y[i] + h*L1],
  L2 = g[t[i] + h, x[i] + h*K1, y[i] + h*L1],
  K3 = f[t[i] + h/2, x[i] + h*K2/2, y[i] + h*L2/2],
  L3 = g[t[i] + h/2, x[i] + h*K2/2, y[i] + h*L2/2],
  K4 = f[t[i] + h, x[i] + h*K3, y[i] + h*L3],
  L4 = g[t[i] + h, x[i] + h*K3, y[i] + h*L3],
  x[i + 1] = x[i] + (K1 + 2 K2 + 2 K3 + K4)*h/6,
  y[i + 1] = y[i] + (L1 + 2 L2 + 2 L3 + L4)*h/6},
  {i, 0, n}]; ListPlot[Table[{x[n], y[n]},
                    {n, 0, 10000}], Frame -> True]
```

Figure 6.5 shows the plot of the fractional Duffing solution from the Example 4.

Example 5 Solve fractional system

$$\begin{cases} D^{1/2}x(t) = 2y, & x(0) = 1, \\ D^{1/2}y(t) = 2z, & y(0) = 1, \\ D^{1/2}z(t) = x - y, & z(0) = 1. \end{cases}$$

Solution

MAPLE

```
> restart;
> Digits:=4:X:=array[0..5]:Y:=array[0..5]:
  Z:=array[0..5]:
```

[7]Ge. Duffing (1861–1944).

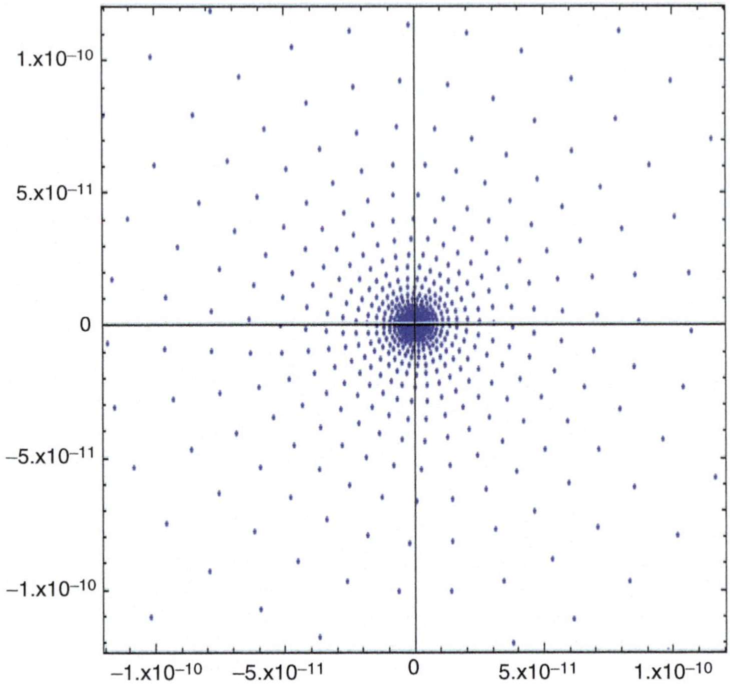

Fig. 6.5 Plot of the fractional Duffing solution from the Example 4

```
> unassign('t,x,y,z'):
> f:=(t,x,y,z)->2*y:
> g:=(t,x,y,z)->2*z:
> p:=(t,x,y,z)->x - y:
> a:=1/2: x:=1:X[0]:=1: y:=1:Y[0]:=1:
> z:=1:Z[0]:=1: d:=0.2:
> h:=evalf(d^a/GAMMA(a+1)):
> for n from 0 to 5 by 1 do
> k1:=h*f(d*n,x,y,z);
> l1:=h*g(d*n,x,y,z):
> m1:=h*p(d*n,x,y,z):
> k2:=h*f(d*n+h,x+k1,y+l1,z+m1):
> l2:=h*g(d*n+h,x+k1,y+l1,z+m1):
> m2:=h*p(d*n+h,x+k1,y+l1,z+m1);
> k3:=h*f(d*n+h/2,x+k2/2,y+l2/2,z+m2/2):
> l3:=h*g(d*n+h/2,x+k2/2,y+l2/2,z+m2/2):
> m3:=h*p(d*n+h/2,x+k2/2,y+l2/2,z+m2/2):
> k4:=h*f(d*n,x+k3,y+l3,z+m3):
> l4:=h*g(d*n,x+h*k3,y+h*l3,z+h*m3):
```

```
> m4:=h*p(d*n,x+h*k3,y+h*l3,z+h*m3):
> x:=x+(k1+2*k2+2*k3+k4)/6:X[n+1]:=x:
> y:=y+(l1+2*l2+2*l3+l4)/6:Y[n+1]:=y:
> z:=z+(m1+2*m2+2*m3+m4)/6:Z[n+1]:=z od:
> for n from 0 to 5 do print(d*n,X[n],Y[n],Z[n]) od;
                            0.,  1,      1,      1
                            0.2, 2.688, 2.030, 1.108
                            0.4, 5.612, 3.458, 1.795
                            0.6, 10.67, 6.102, 3.458
                            0.8, 19.87, 11.26, 6.694
                            1.0, 37.08, 21.11, 12.70
```

MATHEMATICA

```
Clear["`*"]
f[t_, x_, y_, z_] := 2*y;
g[t_, x_, y_, z_] := 2*z;
p[t_, x_, y_, z_] := x - y;
a = 1/2; d = 0.2; x[0] = 1; y[0] = 1; z[0] = 1;
  tmax = 5;
h = (d)^a/Gamma[a + 1];
Do[{k1 = h f[t[n], x[n], y[n], z[n]];
  l1 = h g[n*d, x[n], y[n], z[n]];
  m1 = h p[n*d, x[n], y[n], z[n]];
  k2 = h f[n*d + h/2, x[n] + k1/2, y[n] + l1/2,
  z[n] + m1/2];
  l2 = h g[n*d + h/2, x[n] + k1/2, y[n] + l1/2,
  z[n] + m1/2];
  m2 = h p[n*d + h/2, x[n] + k1/2, y[n] + l1/2,
  z[n] + m1/2];
  k3 = h f[n*d + h/2, x[n] + k2/2, y[n] + l2/2,
  z[n] + m2/2];
  l3 = h g[n*d + h/2, x[n] + k2/2, y[n] + l2/2,
  z[n] + m2/2];
  m3 = h p[n*d + h/2, x[n] + k2/2, y[n] + l2/2,
  z[n] + m2/2];
  k4 = h f[n*d + h, x[n] + k3, y[n] + l3, z[n] + m3];
  l4 = h g[n*d + h, x[n] + k3, y[n] + l3, z[n] + m3];
  m4 = h p[n*d + h, x[n] + k3, y[n] + l3, z[n] + m3];
  x[n + 1] = x[n] + 1/6 *(k1 + 2 k2 + 2 k3 + k4);
  y[n + 1] = y[n] + 1/6* (l1 + 2 l2 + 2 l3 + l4);
  z[n + 1] = z[n] + 1/6* (m1 + 2 m2 + 2 m3 + m4)};
 Print[0.2*n, "  ", x[n], "  ", y[n], "  ", z[n]],
                                        {n, 0, tmax}]
```

Example 6 (Lorenz Attractor with Interpolation Solution) Solve the Lorenz attractor problem:

$$\begin{cases} D^{0.998}x(t) = 10(y - x), & x(0) = 1, \\ D^{0.998}y(t) = 28x - y - xz, & y(0) = 1, \\ D^{0.998}z(t) = xy - \frac{8}{3}z, & z(0) = 1, \end{cases}$$

in Mathematica, using the interpolation command.

```
Clear["`*"]
f[t_, x_, y_, z_] := 10*(y - x);
g[t_, x_, y_, z_] := 28*x - y - x*z;
p[t_, x_, y_, z_] := x*y - 8/3* z;
a = 0.998; d = 0.01;
x[0] = 1;
y[0] = 1;
z[0] = 1;
tmax = 10000;
h = (d)^a/Gamma[a + 1]; Do[{k1 = h f[t[n], x[n],
   y[n], z[n]];
   l1 = h g[n*d, x[n], y[n], z[n]];
   m1 = h p[n*d, x[n], y[n], z[n]];
   k2 = h f[n*d + h/2, x[n] + k1/2, y[n] + l1/2, z[n]
   + m1/2];
   l2 = h g[n*d + h/2, x[n] + k1/2, y[n] + l1/2, z[n]
   + m1/2];
   m2 = h p[n*d + h/2, x[n] + k1/2, y[n] + l1/2, z[n]
   + m1/2];
   k3 = h f[n*d + h/2, x[n] + k2/2, y[n] + l2/2, z[n]
   + m2/2];
   l3 = h g[n*d + h/2, x[n] + k2/2, y[n] + l2/2, z[n]
   + m2/2];
   m3 = h p[n*d + h/2, x[n] + k2/2, y[n] + l2/2, z[n]
   + m2/2];
   k4 = h f[n*d + h, x[n] + k3, y[n] + l3, z[n] + m3];
   l4 = h g[n*d + h, x[n] + k3, y[n] + l3, z[n] + m3];
   m4 = h p[n*d + h, x[n] + k3, y[n] + l3, z[n] + m3];
   x[n + 1] = x[n] + 1/6 (k1 + 2 k2 + 2 k3 + k4);
   y[n + 1] = y[n] + 1/6 (l1 + 2 l2 + 2 l3 + l4);
   z[n + 1] = z[n] + 1/6 (m1 + 2 m2 + 2 m3 + m4);},
   {n,0,tmax}];
sos = Table[{x[t], y[t], z[t]}, {t, 0, tmax}];
ListPointPlot3D[sos, ImageSize -> 300]
```

```
p = Interpolation /@ Transpose@sos;
ParametricPlot3D[Through@p@t, {t, 0, 10000},
  PlotPoints -> 100,
 ColorFunction -> (Hue[#4] &), ImageSize -> 300]
```

Figures 6.6 and 6.7 show the Lorenz system without and with interpolation, respectively.

Example 7 (Rössler Attractor) Solve the fractional Rössler[8] attractor system, using the RK4 method:

$$\begin{cases} D^{0.98}x(t) = -y - z, & x(0) = 1, \\ D^{0.98}y(t) = x + 0.2y, & y(0) = 1, \\ D^{0.98}z(t) = 0.2 + z(x - 8), & z(0) = 1. \end{cases}$$

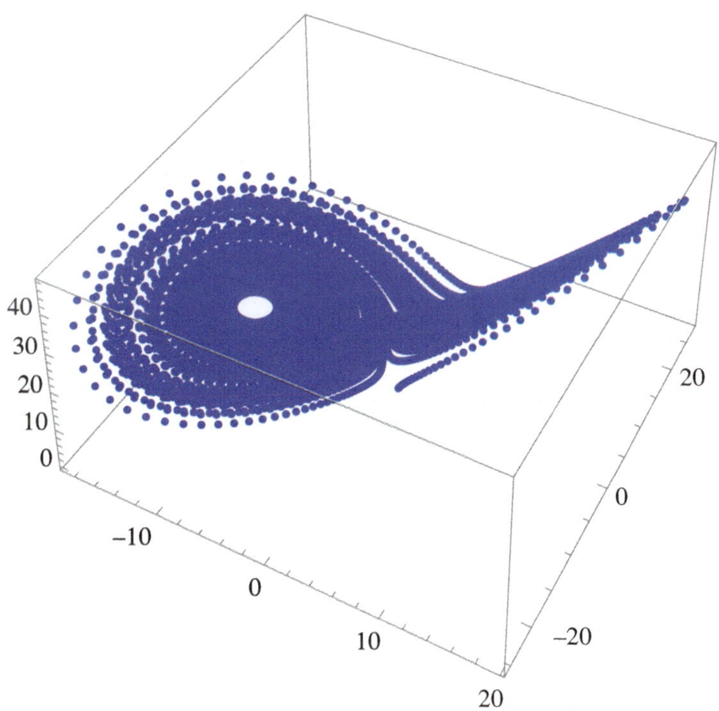

Fig. 6.6 Lorenz system without interpolation

[8]O.E. Rössler(1940–).

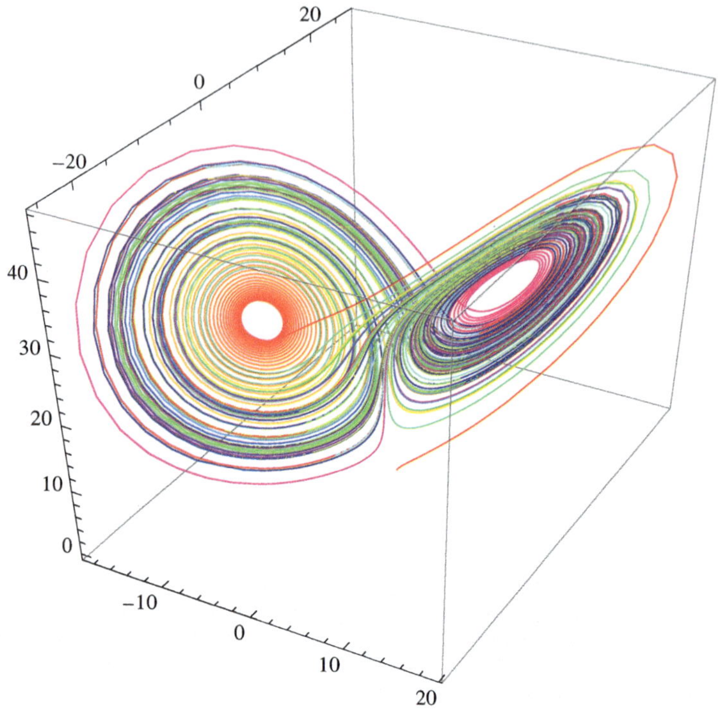

Fig. 6.7 Lorenz system with interpolation

The Mathematica solution is:

```
Clear["`*"]
f[t_, x_, y_, z_] := -y - z;
g[t_, x_, y_, z_] := x + 0.2*y;
p[t_, x_, y_, z_] := 0.2 + z*(x - 8);
a = 0.998; d = 0.01; x[0] = 1;
y[0] = 1; z[0] = 1; tmax = 2000;
h = (d)^a/Gamma[a + 1];
Do[{k1 = h f[t[n], x[n], y[n], z[n]];
  l1 = h g[n*d, x[n], y[n], z[n]];
  m1 = h p[n*d, x[n], y[n], z[n]];
  k2 = h f[n*d + h/2, x[n] + k1/2, y[n] + l1/2,
  z[n] + m1/2];
  l2 = h g[n*d + h/2, x[n] + k1/2, y[n] + l1/2,
  z[n] + m1/2];
  m2 = h p[n*d + h/2, x[n] + k1/2, y[n] + l1/2,
  z[n] + m1/2];
```

```
    k3 = h f[n*d + h/2, x[n] + k2/2, y[n] + l2/2,
    z[n] + m2/2];
    l3 = h g[n*d + h/2, x[n] + k2/2, y[n] + l2/2,
    z[n] + m2/2];
    m3 = h p[n*d + h/2, x[n] + k2/2, y[n] + l2/2,
    z[n] + m2/2];
    k4 = h f[n*d + h, x[n] + k3, y[n] + l3, z[n] + m3];
    l4 = h g[n*d + h, x[n] + k3, y[n] + l3, z[n] + m3];
    m4 = h p[n*d + h, x[n] + k3, y[n] + l3, z[n] + m3];
    x[n + 1] = x[n] + 1/6 (k1 + 2 k2 + 2 k3 + k4);
    y[n + 1] = y[n] + 1/6 (l1 + 2 l2 + 2 l3 + l4);
    z[n + 1] = z[n] + 1/6 (m1 + 2 m2 + 2 m3 + m4);},
                                    {n, 0, tmax}]
ListPointPlot3D[Table[{x[t], y[t], z[t]},
  {t, 0, tmax}],
 ImageSize -> 300]
```

Figure 6.8 shows the 3D Rössler attractor solution $(x(t), y(t), z(t))$.

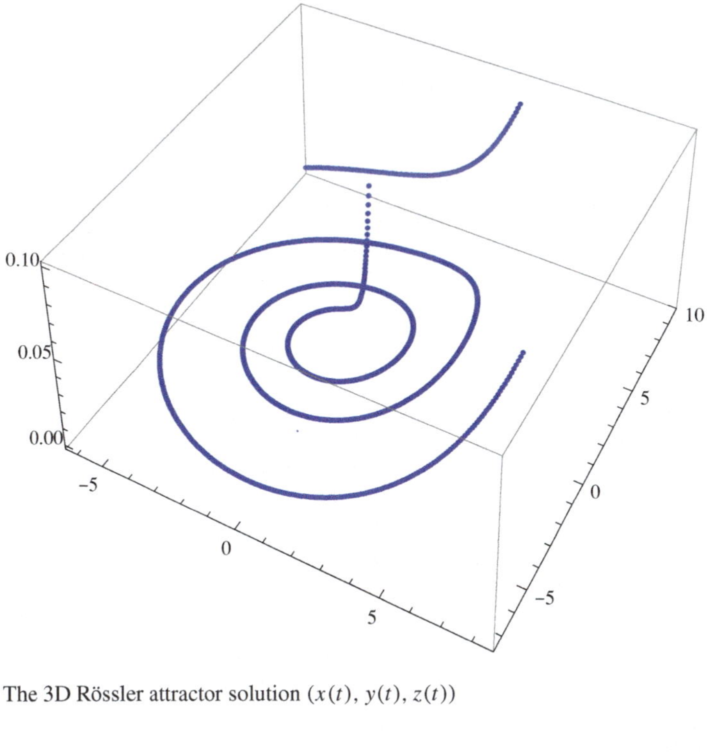

Fig. 6.8 The 3D Rössler attractor solution $(x(t), y(t), z(t))$

Example 8 Find the solution of the fractional Volta attractor:

$$\begin{cases} D^{0.998}x(t) = -x - 5y - yz, & x(0) = 8, \\ D^{0.998}y(t) = -85x - y - xz, & y(0) = 2, \\ D^{0.998}z(t) = 0.5z + xy + 1, & z(0) = 1. \end{cases}$$

Solution in Mathematica, based on the RK4 method.

```
Clear["`*"]
f[t_, x_, y_, z_] := -x - 5*y - z*y;
g[t_, x_, y_, z_] :=  - 85*x - y - x*z;
p[t_, x_, y_, z_] := 0.5*z + x*y + 1;
a = 0.998; d = 0.001; tmax = 10000;
x[0] = 8; y[0] = 2; z[0] = 1;
h = (d)^a/Gamma[a + 1];
Do[{k1 = h f[t[n], x[n], y[n], z[n]];
  l1 = h g[n*d, x[n], y[n], z[n]];
  m1 = h p[n*d, x[n], y[n], z[n]];
  k2 = h f[n*d + h/2, x[n] + k1/2, y[n] + l1/2,
  z[n] + m1/2];
  l2 = h g[n*d + h/2, x[n] + k1/2, y[n] + l1/2,
  z[n] + m1/2];
  m2 = h p[n*d + h/2, x[n] + k1/2, y[n] + l1/2,
  z[n] + m1/2];
  k3 = h f[n*d + h/2, x[n] + k2/2, y[n] + l2/2,
  z[n] + m2/2];
  l3 = h g[n*d + h/2, x[n] + k2/2, y[n] + l2/2,
  z[n] + m2/2];
  m3 = h p[n*d + h/2, x[n] + k2/2, y[n] + l2/2,
  z[n] + m2/2];
  k4 = h f[n*d + h, x[n] + k3, y[n] + l3, z[n] + m3];
  l4 = h g[n*d + h, x[n] + k3, y[n] + l3, z[n] + m3];
  m4 = h p[n*d + h, x[n] + k3, y[n] + l3, z[n] + m3];
  x[n + 1] = x[n] + 1/6 (k1 + 2 k2 + 2 k3 + k4);
  y[n + 1] = y[n] + 1/6 (l1 + 2 l2 + 2 l3 + l4);
  z[n + 1] = z[n] + 1/6 (m1 + 2 m2 + 2 m3 + m4);},
  {n,0,tmax}]
sos = Table[{x[t], y[t], z[t]}, {t, 0, tmax}];
ListPointPlot3D[Table[{x[t], y[t], z[t]},{t,0,tmax}],
 ImageSize -> 300]
 sos = Table[{x[n], y[n], z[n]}, {n, 1, 10000}];
p = Interpolation /@ Transpose@sos;
ParametricPlot3D[Evaluate@Through@p@t,{t,1,10000},
                                 ImageSize->300]
```

Figures 6.9 and 6.10 show the fractional Volta attractor solution without and with interpolation, respectively.

Example 9 (Chua Attractor System) Find the numeric solution of the Chua[9] attractor system:

$$\begin{cases} D^{0.998}x(t) = 40(y - x), \qquad x(0) = 0, \\ D^{0.998}y(t) = (28 - 40)x + 28y - xz, \qquad y(0) = 1, \\ D^{0.998}z(t) = xy - 2z, \qquad z(0) = 0. \end{cases}$$

Solution The solution in Mathematica is:

```
Clear["`*"]
f[t_, x_, y_, z_] := 40*(y - x);
g[t_, x_, y_, z_] := x*(28 - 40) + 28*y - x*z;
```

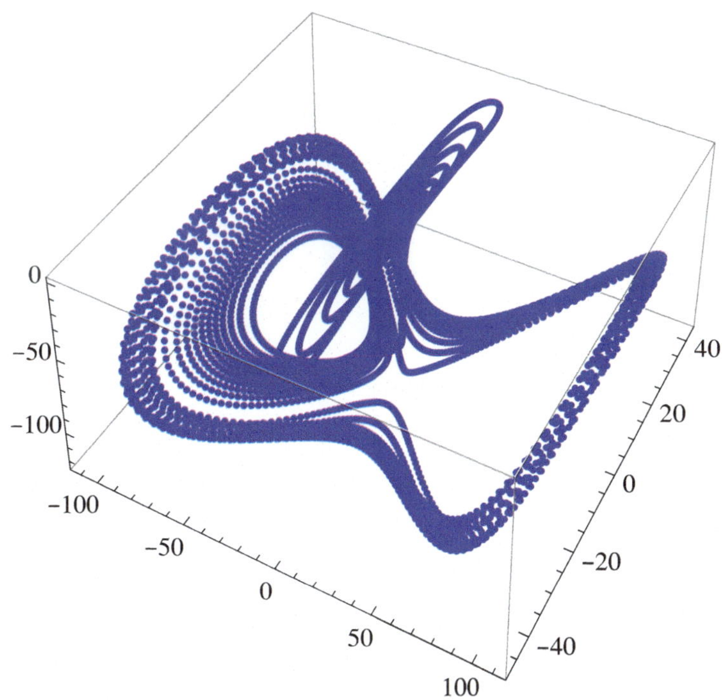

Fig. 6.9 Fractional Volta attractor solution without interpolation

[9]L.O. Chua (1936–).

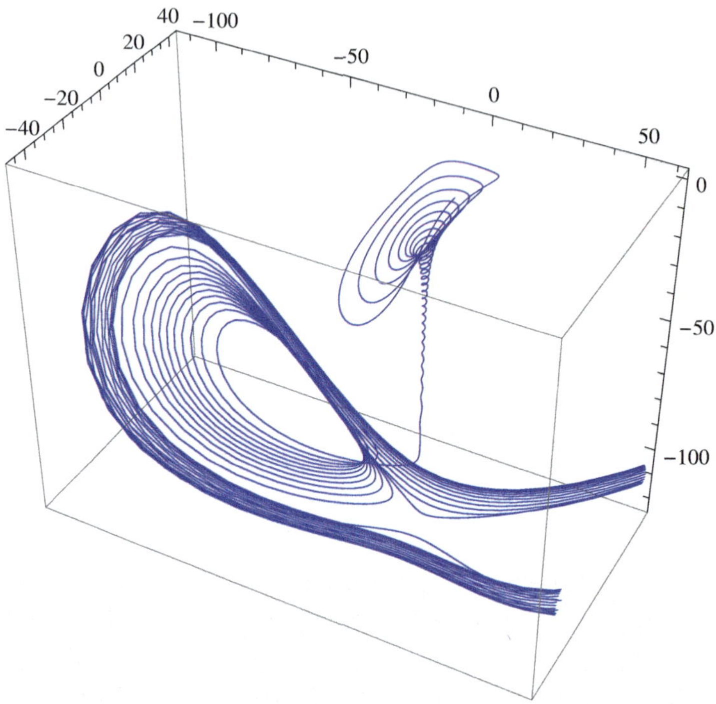

Fig. 6.10 Fractional Volta attractor solution with interpolation

```
p[t_, x_, y_, z_] := x*y - 2* z;
a = 0.998;
d = 0.01;
x[0] = 0;
y[0] = 1;
z[0] = 0;
tmax = 10000;
h = (d)^a/Gamma[a + 1]; Do[{k1 = h f[t[n], x[n],
  y[n], z[n]];
  l1 = h g[n*d, x[n], y[n], z[n]];
  m1 = h p[n*d, x[n], y[n], z[n]];
  k2 = h f[n*d + h/2, x[n] + k1/2, y[n] + l1/2,
  z[n] + m1/2];
  l2 = h g[n*d + h/2, x[n] + k1/2, y[n] + l1/2,
  z[n] + m1/2];
  m2 = h p[n*d + h/2, x[n] + k1/2, y[n] + l1/2,
  z[n] + m1/2];
  k3 = h f[n*d + h/2, x[n] + k2/2, y[n] + l2/2,
  z[n] + m2/2];
```

```
    l3 = h g[n*d + h/2, x[n] + k2/2, y[n] + l2/2,
    z[n] + m2/2];
    m3 = h p[n*d + h/2, x[n] + k2/2, y[n] + l2/2,
    z[n] + m2/2];
    k4 = h f[n*d + h, x[n] + k3, y[n] + l3, z[n] + m3];
    l4 = h g[n*d + h, x[n] + k3, y[n] + l3, z[n] + m3];
    m4 = h p[n*d + h, x[n] + k3, y[n] + l3, z[n] + m3];
    x[n + 1] = x[n] + 1/6 (k1 + 2 k2 + 2 k3 +k4);
    y[n + 1] = y[n] + 1/6 (l1 + 2 l2 + 2 l3 +l4);
    z[n + 1] = z[n] + 1/6 (m1 + 2 m2 + 2 m3 +m4);},
    {n,0,tmax}];
  sos = Table[{x[t], y[t], z[t]}, {t, 0, tmax}];
  ListPointPlot3D[sos, ImageSize -> 300]
  p = Interpolation /@ Transpose@sos;
  ParametricPlot3D[Through@p@t, {t, 0, 10000},
   PlotPoints -> 100,
   ColorFunction -> (Hue[#4] &), ImageSize -> 300]
```

Figures 6.11 and 6.12 show the fractional Chua attractor solution without and with interpolation, respectively.

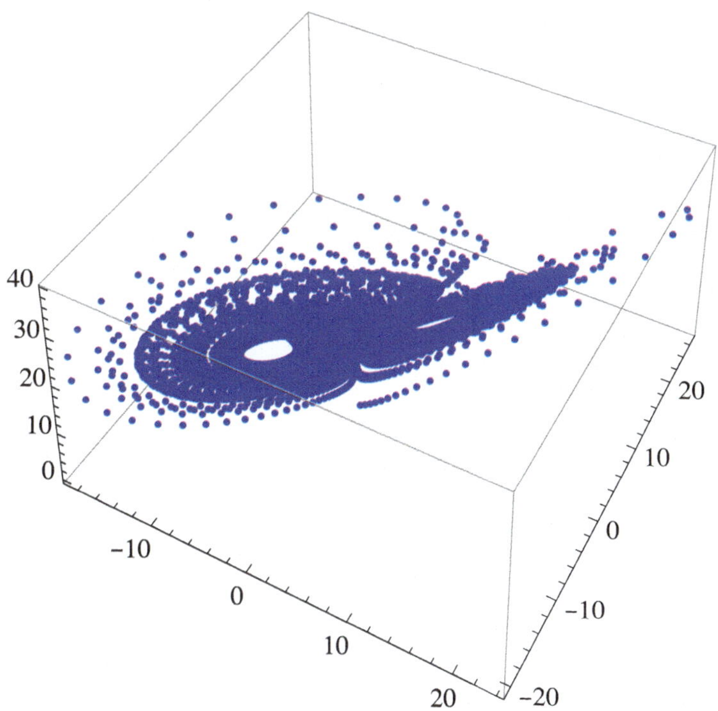

Fig. 6.11 The Chua attractor solution without interpolation

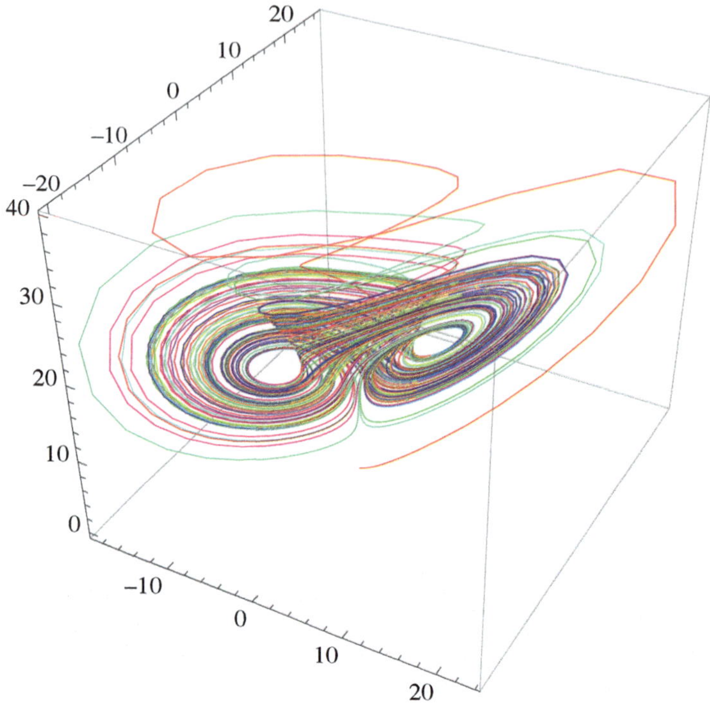

Fig. 6.12 The Chua attractor solution with interpolation

6.5.3 A More General System

Let the following general system of FDE, with initial conditions:

$$\begin{cases} D^{\alpha}x(t) = f(t, x(t), y(t)), & x(t_0) = x_0, \\ D^{\beta}y(t) = g(t, x(t), y(t)), & y(t_0) = y_0, \end{cases}$$

where we suppose that $0 < \beta \leq \alpha \leq 1$.

We will establish here a RK2 approximate solution for this system.

We will denote the temporal step with h, so that the discrete values of time are:

$$t_n = t_0 + nh, \qquad n \in \mathbb{N}.$$

The approximate solution of second order of this system can be written as:

$$x_{n+1} = x_n + \frac{h^{\alpha}}{\Gamma(\alpha + 1)}(K_1 + K_2),$$

$$y_{n+1} = y_n + \frac{h^\beta}{\Gamma(\beta+1)}(L_1 + L_2),$$

where we used the notations:

$$K_1 = f(t_n, x_n, y_n), \qquad L_1 = g(t_n, x_n, y_n),$$

$$K_2 = f\left(t_n + \frac{h^\alpha}{\Gamma(\alpha+1)},\ x_n + \frac{h^\alpha}{\Gamma(\alpha+1)}\frac{K_1}{2},\ y_n + \frac{h^\beta}{\Gamma(\beta+1)}\frac{L_1}{2}\right),$$

$$L_2 = g\left(t_n + \frac{h^\beta}{\Gamma(\beta+1)},\ x_n + \frac{h^\alpha}{\Gamma(\alpha+1)}\frac{K_1}{2},\ y_n + \frac{h^\beta}{\Gamma(\beta+1)}\frac{L_1}{2}\right).$$

Proof In this proof the calculations will be presented schematically.

By analogy with Subsection 4.11.1, we will establish an approximation using the Taylor expansion:

$$x_{n+1} = x_n + c_1\frac{t^\alpha}{\Gamma(\alpha+1)}K_1 + c_2\frac{t^\alpha}{\Gamma(\alpha+1)}K_2$$

$$y_{n+1} = y_n + d_1\frac{t^\alpha}{\Gamma(\alpha+1)}L_1 + d_2\frac{t^\alpha}{\Gamma(\alpha+1)}L_2$$

where

$$K_1 = f(t_n, x_n, y_n),$$

$$L_1 = g(t_n, x_n, y_n),$$

$$K_2 = f\left(t_n + \frac{h^\alpha}{\Gamma(\alpha+1)}a_{f\alpha},\ x_n + \frac{h^\alpha}{\Gamma(\alpha+1)}K_1 b_{f\alpha},\ y_n + \frac{h^\beta}{\Gamma(\beta+1)}L_1 b_{f\beta}\right),$$

$$L_2 = g\left(t_n + \frac{h^\beta}{\Gamma(\beta+1)}a_{g\beta},\ x_n + \frac{h^\alpha}{\Gamma(\alpha+1)}K_1 b_{g\alpha},\ y_n + \frac{h^\beta}{\Gamma(\beta+1)}L_1 b_{g\beta}\right),$$

where the constants c_1, c_2, d_1, d_2, $a_{f\alpha}$, $b_{f\alpha}$, $a_{g\alpha}$, $b_{g\alpha}$, $a_{f\beta}$, $b_{f\beta}$, $a_{g\beta}$, and $b_{g\beta}$ will be established.

We define the errors $E(h)$ and $F(h)$ of expansions

$$E(h) = x(t_n + h) - x(t_n) = \frac{h^\alpha}{\Gamma(\alpha+1)}D^\alpha x(t_n) + \frac{h^{2\alpha}}{\Gamma(2\alpha+1)}D^{2\alpha}x(t_n) + \mathcal{O}(h^{3\alpha}),$$

$$F(h) = y(t_n + h) - y(t_n) = \frac{h^\beta}{\Gamma(\beta+1)}D^\beta y(t_n) + \frac{h^{2\beta}}{\Gamma(2\beta+1)}D^{2\beta}y(t_n) + \mathcal{O}(h^{3\beta}),$$

and the derivatives:

$$D^{\alpha}E(h) = f(t_n, x_n, y_n) + \mathcal{O}(h^{\alpha}) \;\;\Rightarrow\;\; D^{\alpha}E(0) = f(t_n, x_n, y_n),$$

$$D^{\beta}F(h) = g(t_n, x_n, y_n) + \mathcal{O}(h^{\beta}) \;\;\Rightarrow\;\; D^{\beta}E(0) = g(t_n, x_n, y_n).$$

On the other hand

$$D^{\alpha}E(h) = c_1K_1 + c_2K_2 + \mathcal{O}(h\alpha) \;\;\Rightarrow\;\; D^{\alpha}E(0) = (c_1 + c_2)f(t_n, x_n, y_n),$$

$$D^{\beta}E(h) = d_1L_1 + d_2L_2 + \mathcal{O}(h\alpha) \;\;\Rightarrow\;\; D^{\beta}E(0) = (d_1 + d_2)g(t_n, x_n, y_n),$$

hence, by minimization ($h \to 0$) we obtain:

$$\begin{cases} c_1 + c_2 = 1, \\ d_1 + d_2 = 1. \end{cases}$$

The derivative of $E(h)$ is

$$\begin{aligned} D^{2\alpha}E(h) &= D^{\alpha}(D^{\alpha}E(h)) \\ &= D^{\alpha}(c_1K_1 + c_2K_2) + D^{\alpha}\left(c_2\frac{h^{\alpha}}{\Gamma(\alpha+1)}D^{\alpha}K_2\right) + \mathcal{O}(h^{\alpha}), \end{aligned}$$

and for $h \to 0$ we have

$$D^{2\alpha}E(0) = 2c_2D^{2\alpha}K_2(0).$$

But:

$$D^{\alpha}f(t, x(t), y(t)) = f_t + f_xD^{\alpha}x + f_yD^{\alpha}y = f_t + f_xf,$$

$$D^{\alpha}y = 0 \quad \text{because: } \alpha > \beta,$$

we obtain

$$\begin{cases} 2c_2a_{f\alpha} = 1, \\ 2c_2b_{f\alpha 1} = 1, \\ 2c_2b_{f\alpha 2} = 1. \end{cases}$$

The calculations will continue with the derivative. Similar computations can be carried out in the case of RK4 method.

It results:

$$x_{n+1} = x_n + \frac{1}{6}\frac{h^{\alpha}}{\Gamma(\alpha+1)}(K_1 + 2K_2 + 2K_3 + K_4),$$

$$y_{n+1} = y_n + \frac{1}{6}\frac{h^{\beta}}{\Gamma(\beta+1)}(L_1 + 2L_2 + 2L_3 + L_4).$$

Example 1 (Modified Duffing System) Find the RK4 solution for the modified fractional Duffing system:

$$\begin{cases} D^{0.998}x(t) = y, \qquad x(0) = 1, \\ D^{0.50}y(t) = -x + 0.25(1 - x^3)y, \qquad y(0) = 0. \end{cases}$$

The RK4 solution in Mathematica is:

```
Clear["`*"];
a = 0.998; b = 0.5; x[0] = 1; y[0] = 0;
h = (0.2)^a/Gamma[a + 1]; l = (0.2)^b/Gamma[b + 1];
n = 10000;
Do[t[i] = 0.0 + (0.2)*i, {i, 0, n}]
f[t_, x_, y_] := y
g[t_, x_, y_] := -x + 0.25*(1 - x^2)*y
Do[{K1 = f[t[i], x[i], y[i]], L1 = g[t[i], x[i],
  y[i]],
  K2 = f[t[i] + h, x[i] + h*K1, y[i] + h*L1],
  L2 = g[t[i] + l, x[i] + l*K1, y[i] + l*L1],
  K3 = f[t[i] + h/2, x[i] + h*K2/2, y[i] + h*L2/2],
  L3 = g[t[i] + l/2, x[i] + l*K2/2, y[i] + l*L2/2],
  K4 = f[t[i] + h, x[i] + h*K3, y[i] + h*L3],
  L4 = g[t[i] + l, x[i] + l*K3, y[i] + l*L3],
  x[i + 1] = x[i] + (K1 + 2 K2 + 2 K3 + K4)*h/6,
  y[i + 1] = y[i] + (L1 + 2 L2 + 2 L3 + L4)*l/6},
  {i,0,n}];
  ListPlot[Table[{x[n], y[n]}, {n, 0, 10000}],
                                  Frame -> True]
sol = Table[{x[n], y[n]}, {n, 1, 1000}];
p = Interpolation /@ Transpose@sol;
ParametricPlot[Evaluate@Through@p@t, {t, 1, 1000},
                                     Frame -> True]
```

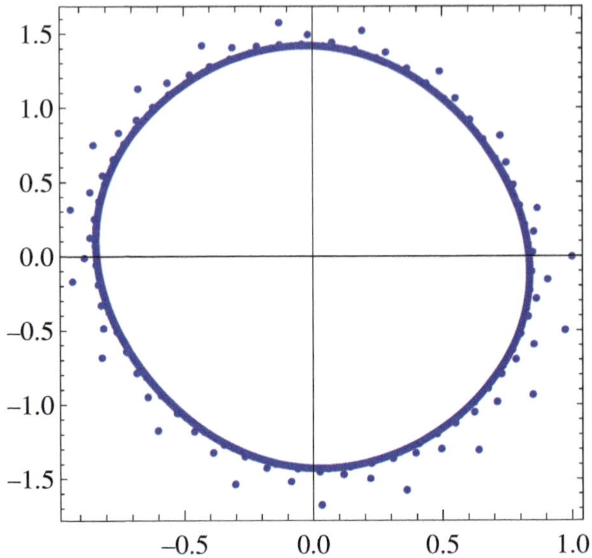

Fig. 6.13 RK4 solution $(x(t), y(t))$ of the modified fractional Duffing system without iteration

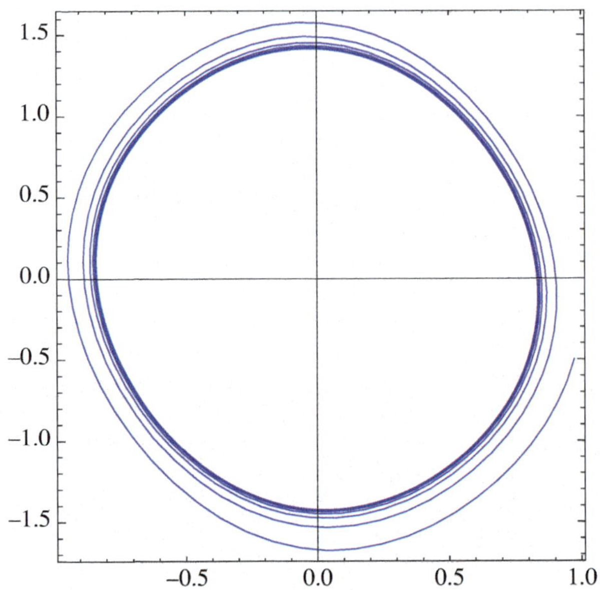

Fig. 6.14 RK4 solution $(x(t), y(t))$ of the modified fractional Duffing system with iteration

Figures 6.13 and 6.14 show the RK4 solution $(x(t), y(t))$ of the modified fractional Duffing system without and with iteration, respectively.

Remark Starting from the fractional RK procedures introduced in this section, the reader can establish similar procedures for the other RK methods, established for the integer order cases [1].

The Fractional Colpitts Oscillator

Solve the fractional Colpitts[10] oscillator described by the system of FDE:

$$\begin{cases} D^{0.998}\, x(t) = y, & x(0) = 0, \\ D^{0.998}\, y(t) = z, & y(0) = 0.4, \\ D^{0.998}\, z(t) = -z - x - 10^{-9}(\exp(y) - 1), & z(0) = 0, \end{cases}$$

using RK4 method in Mathematica.

Solution

```
Clear["`*"]
f[t_, x_, y_, z_] := y;
g[t_, x_, y_, z_] := z;
p[t_, x_, y_, z_] := -z - x - 10^(-9) (Exp[y] - 1);
a = 0.998;
d = 0.01;
x[0] = 0;
y[0] = 0.4;
z[0] = 0;
tmax = 10000;
h = (d)^a/Gamma[a + 1];
Do[{k1 = h f[t[n], x[n], y[n], z[n]];
  l1 = h g[n*d, x[n], y[n], z[n]];
  m1 = h p[n*d, x[n], y[n], z[n]];
  k2 = h f[n*d + h/2, x[n] + k1/2, y[n] + l1/2,
  z[n] + m1/2];
  l2 = h g[n*d + h/2, x[n] + k1/2, y[n] + l1/2,
  z[n] + m1/2];
  m2 = h p[n*d + h/2, x[n] + k1/2, y[n] + l1/2,
  z[n] + m1/2];

  k3 = h f[n*d + h/2, x[n] + k2/2, y[n] + l2/2,
  z[n] + m2/2];
  l3 = h g[n*d + h/2, x[n] + k2/2, y[n] + l2/2,
  z[n] + m2/2];
  m3 = h p[n*d + h/2, x[n] + k2/2, y[n] + l2/2,
  z[n] + m2/2];
  k4 = h f[n*d + h, x[n] + k3, y[n] + l3, z[n] + m3];
  l4 = h g[n*d + h, x[n] + k3, y[n] + l3, z[n] + m3];
  m4 = h p[n*d + h, x[n] + k3, y[n] + l3, z[n] + m3];
  x[n + 1] = x[n] + 1/6 (k1 + 2 k2 + 2 k3 + k4);
```

[10] E.H. Colpitts (1872–1949).

```
    y[n + 1] = y[n] + 1/6 (l1 + 2 l2 + 2 l3 + l4);
    z[n + 1] = z[n] + 1/6 (m1 + 2 m2 + 2 m3 + m4);},
                                      {n, 0, tmax}]
ListPointPlot3D[Table[{x[t], y[t], z[t]},
     {t, 0, tmax}], ImageSize -> 300]
 sos = Table[{x[t], y[t], z[t]}, {t, 0, tmax}];
p = Interpolation /@ Transpose@sos;
ParametricPlot3D[Through@p@t, {t, 0, 10000},
PlotPoints -> 100,
 ColorFunction -> (Hue[#4] &), ImageSize -> 300]
```

Figures 6.15 and 6.16 show the RK4 solution $(x(t), y(t))$ of the fractional Colpitts oscillator system without and with iteration, respectively.

The Fractional Sprott Oscillator

Solve the fractional Sprott[11] oscillator described by the system of FDE:

$$\begin{cases} D^{0.998}\, x(t) = y, & x(0) = -0.5, \\ D^{0.998}\, y(t) = z, & y(0) = 0, \\ D^{0.998}\, z(t) = -z - x - 10^{-9}(\exp(y) - 1), & z(0) = 0, \end{cases}$$

using RK4 method in Mathematica.

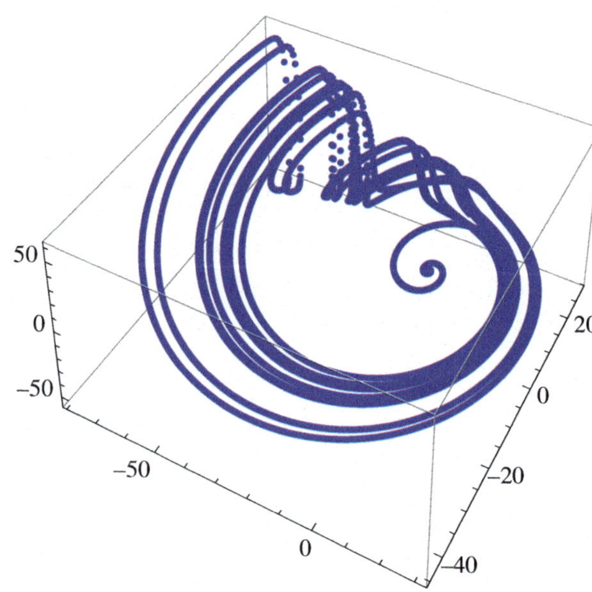

Fig. 6.15 RK4 solution of the fractional Colpitts oscillator system without iteration

[11]D.A. Sprott (1930–2013).

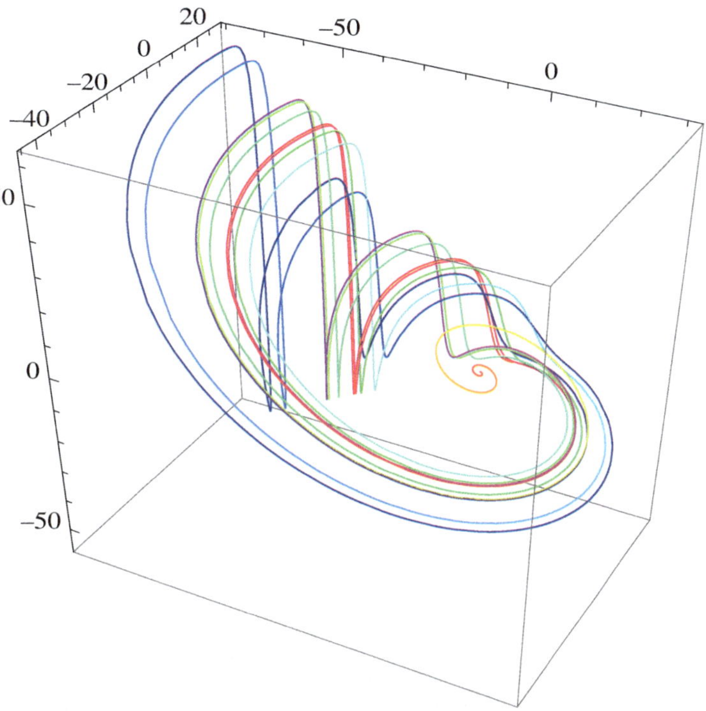

Fig. 6.16 RK4 solution of the fractional Colpitts oscillator system with iteration

Solution

```
Clear["`*"]
f[t_, x_, y_, z_] := y;
g[t_, x_, y_, z_] := z;
p[t_, x_, y_, z_] := -x - y - Sign[1 + 4*y];
a = 0.998; d = 0.01; x[0] = -0.5; y[0] = 0;
z[0] = 0; tmax = 2000; h = (d)^a/Gamma[a + 1];
Do[{k1 = h f[t[n], x[n], y[n], z[n]];
  l1 = h g[n*d, x[n], y[n], z[n]];
  m1 = h p[n*d, x[n], y[n], z[n]];
  k2 = h f[n*d + h/2, x[n] + k1/2, y[n] + l1/2,
  z[n] + m1/2];
  l2 = h g[n*d + h/2, x[n] + k1/2, y[n] + l1/2,
  z[n] + m1/2];
  m2 = h p[n*d + h/2, x[n] + k1/2, y[n] + l1/2,
  z[n] + m1/2];
```

```
k3 = h f[n*d + h/2, x[n] + k2/2, y[n] + l2/2,
z[n] + m2/2];
l3 = h g[n*d + h/2, x[n] + k2/2, y[n] + l2/2,
z[n] + m2/2];
m3 = h p[n*d + h/2, x[n] + k2/2, y[n] + l2/2,
z[n] + m2/2];
k4 = h f[n*d + h, x[n] + k3, y[n] + l3, z[n] + m3];
l4 = h g[n*d + h, x[n] + k3, y[n] + l3, z[n] + m3];
m4 = h p[n*d + h, x[n] + k3, y[n] + l3, z[n] + m3];
x[n + 1] = x[n] + 1/6 (k1 + 2 k2 + 2 k3 + k4);
y[n + 1] = y[n] + 1/6 (l1 + 2 l2 + 2 l3 + l4);
z[n + 1] = z[n] + 1/6 (m1 + 2 m2 + 2 m3 + m4);},
  {n, 0, tmax}]
ListPointPlot3D[Table[{x[t], y[t], z[t]},
   {t, 0, tmax}], ImageSize -> 300]
    sos = Table[{x[t], y[t], z[t]}, {t, 0, tmax}];
p = Interpolation /@ Transpose@sos                    ;
ParametricPlot3D[Through@p@t, {t, 0, 2000},
 PlotPoints -> 100,
 ColorFunction -> (Hue[#4] &), ImageSize -> 300]
```

Figures 6.17 and 6.18 show the RK4 solution of the fractional Sprott oscillator system without and with iteration, respectively.

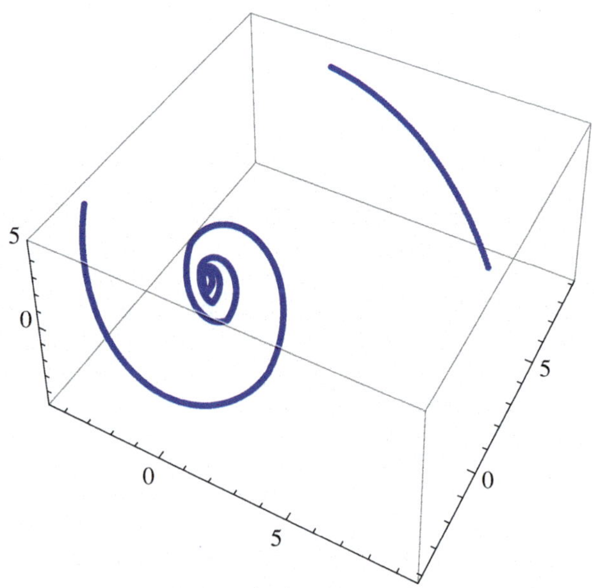

Fig. 6.17 RK4 solution of the Sprott oscillator system without iteration

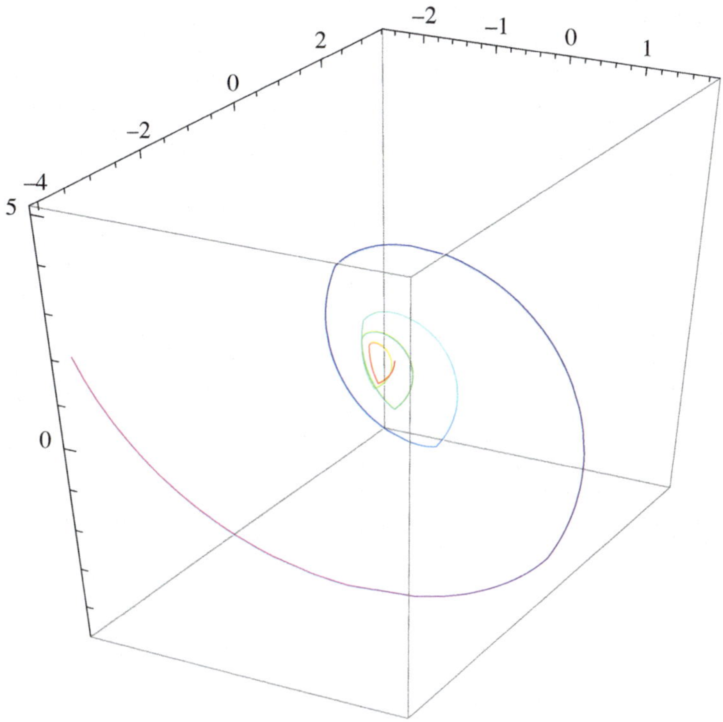

Fig. 6.18 RK4 solution of the Sprott oscillator system with iteration

6.5.4 A Vectorial Runge–Kutta Algorithm

We illustrate the vectorial Runge–Kutta algorithm by means of four examples:

Example 1 (Van Der Pol Fractional Equation)

$$\begin{cases} D^{\alpha}x_1(t) = x_2, & x_1(0) = 0, \\ D^{\alpha}x_2(t) = -x_1 + 2\,(1 - x_1^2)\,x_2, & x_2(0) = 1, \end{cases}$$

for $\alpha = 0.998$.

Solution In Mathematica the solution will be:

```
Clear["`*"]
a = 0.998
RK4[f_, x0_, A_, B_, n_] :=
```

```
Module[{h, K1, K2, K3, K4, Sol = {{A, x0}},
x = x0, t}, h = (B - A)/n;
 Do[t = A + k h^a/Gamma[a + 1];
  K1 = h^a/Gamma[a + 1] f[t, x];
  K2 = h^a/Gamma[a + 1] f[t + (1/2) h^a/Gamma[a + 1],
                                      x + (1/2) K1];
  K3 = h^a/Gamma[a + 1] f[t + (1/2) h^a/Gamma[a + 1],
                                      x + (1/2) K2];
  K4 = h^a/Gamma[a + 1] f[t + h^a/Gamma[a + 1],
  x + K3];
  x = x + (1/6) K1 + (1/3) K2 + (1/3) K3 + (1/6) K4;
  Sol = Append[Sol, {t, x}], {k, 1, n}];
 Sol]
 F[t_, x_] := { x[[2]], 2 (1 - x[[1]] ^2) x[[2]]
 - x[[1]]};

Solution = RK4[F, {0, 1}, 0.0, 100.0, 5000];
A = 0
ListPlot[Take[Solution[[All, 2]], 5000],
 ImageSize -> 300,
 PlotStyle -> {Blue}]
```

Figure 6.19 shows the $(x_1(t), x_2(t))$ vector solution of the Van der Pol system.

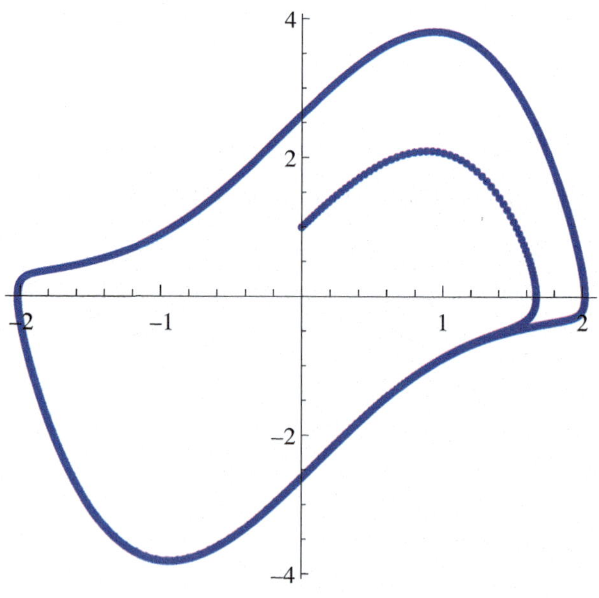

Fig. 6.19 The $(x_1(t), x_2(t))$ vector solution of the Van der Pol system

Example 2 Solve the fractional Rössler attractor system, using the vector RK4 method:

$$\begin{cases} D^{0.98}x_1(t) = -x_2 - x_3, & x_1(0) = 0, \\ D^{0.98}x_2(t) = x_1 + 0.2x_2, & x_2(0) = 1, \\ D^{0.98}x_3(t) = 0.2 + x_3(x_1 - 8), & x_3(0) = 0. \end{cases}$$

Solution

```
Clear["`*"]
a = 0.998
RK4[f_, x0_, A_, B_, n_] :=
 Module[{h, K1, K2, K3, K4, Sol = {{A, x0}},
 x = x0, t}, h = (B - A)/n;
  Do[t = A + k h^a/Gamma[a + 1];
   K1 = h^a/Gamma[a + 1] f[t, x];
   K2 = h^a/Gamma[a + 1] f[t +
            (1/2) h^a/Gamma[a + 1], x + (1/2) K1];
   K3 = h^a/Gamma[a + 1] f[t +
            (1/2) h^a/Gamma[a + 1], x + (1/2) K2];
   K4 = h^a/Gamma[a + 1] f[t +
                 h^a/Gamma[a + 1], x + K3];
   x = x + (1/6) K1 + (1/3) K2 +
       (1/3) K3 + (1/6) K4;
   Sol = Append[Sol, {t, x}], {k, 1, n}];
  Sol]
  F[t_, x_] := {-x[[2]] - x[[3]], x[[1]]
    + 0.2 x[[2]], 0.2 + x[[1]] x[[3]] -
                             5.7 x[[3]]};

Solution = RK4[F, {0, 1.0, 0}, 0.0, 200.0, 5000];
A = 0
ListPointPlot3D[Take[Solution[[All, 2]], 1000],
                                  PlotStyle -> {Blue}]
```

Figure 6.20 shows the $(x_1(t), x_2(t))$ vector solution of the Rössler attractor.

Example 3 (Volta Fractional Attractor) Find the solution of the fractional Volta attractor using the vector RK4 algorithm:

$$\begin{cases} D^{0.998}x_1(t) = -x_1 - 5x_2 - x_2x_3, & x_1(0) = 8, \\ D^{0.998}x_2(t) = -85x_1 - x_2 - x_1x_3, & x_2(0) = 2, \\ D^{0.998}x_3(t) = 0.5x_3 + x_1x_2 + 1, & x_3(0) = 1. \end{cases}$$

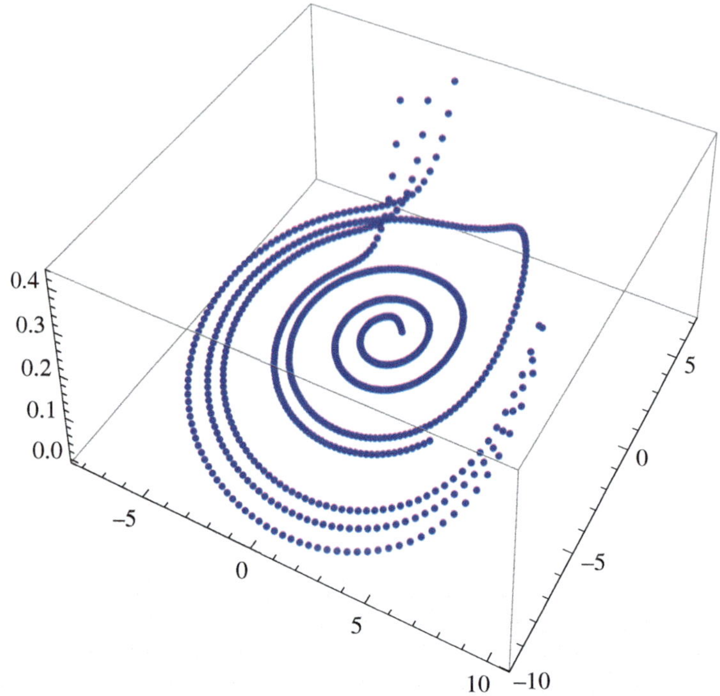

Fig. 6.20 The $(x_1(t), x_2(t), x_3(t))$ vector solution of the Rössler attractor

Solution

In Mathematica can be written:

```
Clear["`*"]
a = 0.998
RK4[f_, x0_, A_, B_, n_] :=
 Module[{h, K1, K2, K3, K4, Sol = {{A, x0}},
  x = x0, t}, h = (B - A)/n;
  Do[t = A + k h^a/Gamma[a + 1];
   K1 = h^a/Gamma[a + 1] f[t, x];
   K2 = h^a/Gamma[a + 1] f[t + (1/2) h^a/Gamma[a + 1],
                                    x + (1/2) K1];
   K3 = h^a/Gamma[a + 1] f[t + (1/2) h^a/Gamma[a + 1],
                                    x + (1/2) K2];
   K4 = h^a/Gamma[a + 1] f[t + h^a/Gamma[a + 1],
   x + K3];
   x = x + (1/6) K1 + (1/3) K2 + (1/3) K3 + (1/6) K4;
   Sol = Append[Sol, {t, x}], {k, 1, n}];
  Sol]
```

```
F[t_, x_] := {-x[[1]] - 5 x[[2]] - x[[2]] x[[3]],
  -85 x[[1]]
   - x[[2]] - x[[1]] x[[3]], x[[1]] x[[2]] + 0.5
   x[[3]] + 1};

Solution = RK4[F, {8.0, 2.0, 1.0}, 0.0, 100.0, 5000];
A = 0
ListPointPlot3D[Take[Solution[[All, 2]], 1000],
                                    PlotStyle -> {Blue}]
```

Figure 6.21 shows the $(x_1(t), x_2(t))$ vector solution of the Volta attractor.

Example 4 (Lorenz Fractional Attractor) Solve the Lorentz attractor system:

$$\begin{cases} D^{0.98}x_1(t) = -10(x_1 - x_2), & x_1(0) = 1, \\ D^{0.98}x_2(t) = 28x_1 - x_2 - x_1x_3, & x_2(0) = 1, \\ D^{0.98}x_3(t) = x_1x_2 - \frac{8}{3}x_3, & x_3(0) = 1. \end{cases}$$

using the vector RK4 method, in Mathematica.

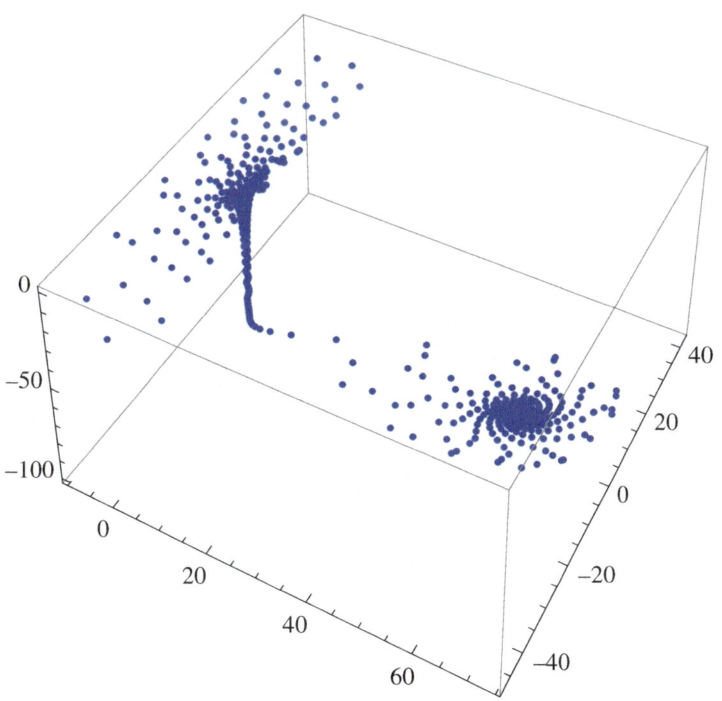

Fig. 6.21 The $(x_1(t), x_2(t), x_3(t))$ vector solution of the Volta attractor

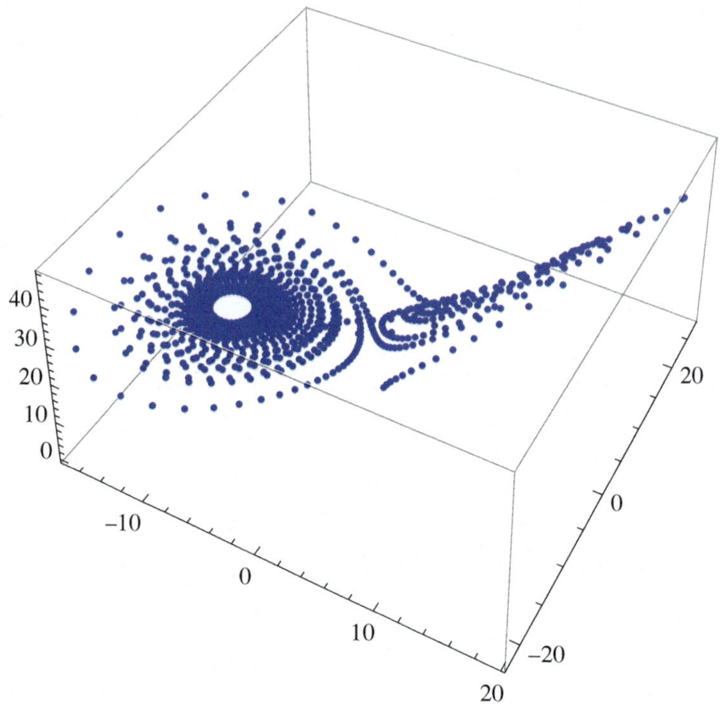

Fig. 6.22 The $(x_1(t), x_2(t), x_3(t))$ vector solution of the Lorenz attractor

Solution

```
Clear["`*"]
a = 0.998
RK4[f_, x0_, A_, B_, n_] :=
 Module[{h, K1, K2, K3, K4, Sol = {{A, x0}},
 x = x0, t}, h = (B - A)/n;
  Do[t = A + k h^a/Gamma[a + 1];
   K1 = h^a/Gamma[a + 1] f[t, x];
   K2 = h^a/Gamma[a + 1] f[t + (1/2) h^a/Gamma[a + 1],
                                 x + (1/2) K1];
   K3 = h^a/Gamma[a + 1] f[t + (1/2) h^a/Gamma[a + 1],
                                 x + (1/2) K2];
   K4 = h^a/Gamma[a + 1] f[t + h^a/Gamma[a + 1],
   x + K3];
   x = x + (1/6) K1 + (1/3) K2 + (1/3) K3 + (1/6) K4;
   Sol = Append[Sol, {t, x}], {k, 1, n}];
  Sol]
F[t_, x_] := {10 (x[[2]] - x[[1]]), x[[1]]
  (28 - x[[3]])
```

```
    - x[[2]], x[[1]] x[[2]] - 8/3 x[[3]]};

Solution = RK4[F, {1.0, 1.0, 1.0}, 0.0, 100.0, 5000];
A = 0
ListPointPlot3D[Take[Solution[[All, 2]], 1000],
                                    PlotStyle -> {Blue}]
```

Figure 6.22 shows the $(x_1(t), x_2(t))$ vector solution of the Lorenz attractor.

References

1. Butcher, J. C. (1987). *The numerical analysis of ordinary differential equations, Runge-Kutta and general linear methods*. New York: Wiley-Interscience.
2. Drăgănescu, G. E. (2006). Application of a variational iteration method to linear and nonlinear viscoelastic models with fractional derivatives. *Journal of Mathematical Physics, 47*, 082902.
3. He, J.-H. (1999). Variational iteration method - a kind of non-linear analytical technique: Some examples. *International Journal of Non-linear Mechanics, 34*, 699–708.
4. He, J.-H. (2006). Some asymptotic methods for strongly nonlinear equations. *International Journal of Modern Physics B, 20*, 1141–1199.
5. Wu, G.-C. (2011). A fractional variational iteration method for solving fractional nonlinear differential equations. *Computers & Mathematics with Applications, 61*, 2186–2190.
6. Wu, G.-C., & Baleanu, D. (2013). Variational iteration method for fractional calculus - a universal Laplace transform. *Advances in Difference Equations, 18*, 1–9.

Index

C. Milici et al., *Introduction to Fractional Differential Equations*, Nonlinear Systems and Complexity 25, https://doi.org/10.1007/978-3-030-00895-6

Printed by Printforce, the Netherlands